UK Airfields of the Cold War

UK Airfields of the
Cold War

Philip Birtles

MIDLAND

UK Airfields of the Cold War
Philip Birtles

First published 2012

ISBN 978 1 85780 346 4

All rights reserved. No part of this book may be reproduced or transmitted in any form or by any means, electronic or mechanical, including copying, recording, scanning or by any information storage and retrieval system, on the internet or elsewhere, without permission from the Publisher in writing.

© Philip Birtles 2012

Published by Midland Publishing

an imprint of Ian Allan Publishing Ltd, Hersham, Surrey KT12 4RG.
Printed in England by Ian Allan Printing Ltd, Hersham, Surrey KT12 4RG.

Visit the Ian Allan Publishing website at www.ianallanpublishing.com

Distributed in the United States of America and Canada by BookMasters Distribution Services.

Copyright
Illegal copying and selling of publications deprives authors, publishers and booksellers of income, without which there would be no investment in new publications. Unauthorised versions of publications are also likely to be inferior in quality and contain incorrect information. You can help by reporting copyright infringements and acts of piracy to the Publisher or the UK Copyright Service.

Contents

	Introduction	7
Chapter 1	The Berlin Airlift	9
Chapter 2	The Airfields	23
Chapter 3	Tactical RAF Fighter Stations	27
Chapter 4	RAF Tactical Bomber Stations	61
Chapter 5	Britain's USAF Airfields	77
Chapter 6	RAF Strategic Fighter Stations	114
Chapter 7	RAF Strategic Bomber Stations	124
Chapter 8	Maritime Reconnaissance	153
	Bibliography	160

Abbreviations used in this book

AAF	Army Air Force
AAM	Air-to-Air Missile
ADR	Air Defence Region (UK)
A&AEE	Aeroplane & Armament Experimental Establishment
AEW	Airborne Early Warning
AF	Air Force
AFB	Air Force Base
AFDS	Air Fighting Development Squadron
AFDU	Air Fighting Development Unit
AFNORTH	Allied Forces, Northern Europe
AFVG	Anglo French Variable Geometry
ANG	Air National Guard
AP/ANS	Advance Programme/Air Navigation System
APC	Armament Practice Camp
AS	Aggressor Squadron
ASARS	Advanced Synthetic Aperture Radar System
ASP	Aircraft Servicing Platform
ASR	Air-Sea Rescue
ASV	Air-To-Surface Vessel
ASWDU	Air-Sea Warfare Development Unit
AWDS	All Weather Development Squadron
AWACS	Airborne Warning & Control System
BAC	British Aircraft Corporation
BCDU	Bomber Command Development Unit
BAe	British Aerospace
BG	Bomber Group
BMEWS	Ballistic Missile Early Warning System
BNSDS	Bombing & Navigation Systems Development Squadron
BOAC	British Overseas Airways Corporation
BS	Bomber Squadron
BW	Bomber Wing
C&M	Care and Maintenance
CAS	Close Air Support
CFE	Central Fighter Establishment
CFS	Central Flying School
CIA	Central Intelligence Agency (US)
CND	Campaign for Nuclear Disarmament
CPS	Central Processing System
DACT	Dissimilar Air Combat Training
DEFCON	Defence Readiness Condition
DFCS	Defence Fire Control System
DFLS	Day Fighter Leaders School
DOB	Dispersal Operating Base
DOD	Department of Defense
ECS	Electronic Countermeasures Squadron
ECW	Electronic Countermeasures Wing
ESM	Electronic Surveillance Measures
EW	Early Warning
EXCOMM	Executive Committee of the National Security Council
FAR	Fuerza Aerea Revolucionaria
FAA	Fleet Air Arm
FBW	Fighter Bomber Wing
FIDO	Fog Investigation & Dispersal Operation
FIW	Fighter Interception Wing
FLIR	Forward Looking Infrared
Flt	Flight
FOL	Forward Operating Location
FS	Fighter Squadron
FTS	Flying Training School
GAMA	GLCM Alert & Maintenance Area
GCA	Ground Control Approach
GLCM	Ground Launched Cruise Missiles
GPS	Global Positioning System
GSU	Group Standardisation Unit
HCT	Harrier Conversion Team
HCU	Harrier Conversion Unit
HGCU	Heavy Glider Conversion Unit
HMS	His (or Her) Majesty's Ship
HQ	Headquarters
HRH	His (or Her) Royal Highness
HSA	Hawker Siddeley Aviation
HUD	Head-Up Display
ICBM	Intercontinental Ballistic Missile
IED	Improvised Explosive Device
IR	Infra Red
IRBM	Intermediate-Range Ballistic Missile
ITR	Information Technology Research
IWM	Imperial War Museum
JFH	Joint Force Harrier
KGB	Komitet Gosudarstvennoy Bezopasnosti (USSR State Security)
KLM	Koninklijke Luchtvaart Maatschappij (Dutch Royal Airlines)
LABS	Low Approach Bombing System
LCC	Launch Control Centres
MAD	Magnetic Anomaly Detector
MAP	Military Aid Programme
MATS	Military Air Transport Service
METS	Multi-Engine Training Squadron
MBB	Messerschmitt-Bolkow-Blohm
MIRV	Multiple Independently-Targetable Re-entry Vehicle
MLU	Mid-Life Update
MOD	Ministry of Defence
MOS	Ministry of Supply
MOTU	Maritime Operational Training Unit
MR	Maritime Reconnaissance
MRCA	Multi-Role Combat Aircraft
MRBM	Medium-range Ballistic Missiles
MRR	Maritime Radar Reconnaissance
MU	Maintenance Unit
NAMMO	NATO MRCA Management Organisation
NAS	Naval Air Squadron
NASA	National Air & Space Agency
NATO	North Atlantic Treaty Organisation
NBSDU	Navigational Bomb Sight Development Unit
NFLS	Night Fighter Leaders School
NORAD	North American Aerospace Defense Command
OCU	Operational Conversion Unit
ORP	Operation Readiness Platform
PLSS	Precision Locator Strike System
PR	Photographic Reconnaissance
PRU	Photographic Reconnaissance Unit
QRA	Quick Reaction Alert
RAF	Royal Air Force
RAAF	Royal Australian Air Force
RAuxAF	Royal Auxiliary Air Force
RATO	Rocket Assisted Take-Off
RCAF	Royal Canadian Air Force
RIAT	Royal International Air Tattoo
RN	Royal Navy
RNAS	Royal Naval Air Service
ROC	Royal Observer Corps
RS	Reconnaissance Squadron
RW	Reconnaissance Wing
SAC	Strategic Air Command
SACEUR	Supreme Allied Commander Europe
SACLANT	Supreme Allied Commander Atlantic
SAM	Surface to Air Missile
SDSR	Strategic Defence Services Review
SEAD	Suppression of Enemy Air Defences
SFTS	Synthetic Flight Training Systems
SIGINT	Signals Intelligence
SLAR	Side Looking Airborne Radar
SLIR	Side Looking Infra Red
SLG	Satellite Landing Ground
SRS	Strategic Reconnaissance Squadron
SRW	Strategic Reconnaissance Wing
SoTT	School of Technical Training
Sqn	Squadron
TCDU	Transport Command Development Unit
TCG	Tactical Communications Group
TCW	Tactical Control Wing
TDY	Temporary Duty Postings
TEL	Transport Erector Launcher
TEREC	Tactical Electronic Reconnaissance
TERCOM	Terrain Contour Matching
TES	Tripartite Evaluation Squadron
TFW	Tactical Fighter Wing
TMW	Tactical Missile Wing
TRS	Tactical Reconnaissance Squadron
TRW	Tactical Reconnaissance Wing
TTTE	Tri-National Tornado-Training Establishment
TWCU	Tornado Weapons Conversion Unit
UK	United Kingdom
UKWMO	United Kingdom Warning & Monitoring Organisation
UN	United Nations
US / USA	United States / United States of America
USAF	United States Air Force
USAAF	United States Army Air Force
USAFE	United States Air Forces in Europe
USSR	Union of Soviet Socialist Republics
USN	United States Navy
V/STOL	Vertical/Short Take Off and Landing
WCU	Washington Conversion Unit
WOD	Wind Over Deck
WW1	World War 1
WW2	World War 2

Introduction

Although this book is primarily about Britain's Cold War airfields, the Allied aircraft serving at these bases cannot be ignored. At the start of the Cold War the RAF was equipped with Meteor and Vampire first-generation jet fighters, each armed with four World War 2-vintage 20mm cannons. The strategy was to put as many of these fighters in the air from east coast airfields to hopefully overwhelm the approaching hostile armada of Soviet bombers, although the altitude and climb performance of these aircraft left a lot to be desired. Fortunately the Soviet threat was from World War 2-vintage technology, but it is hard to imagine that some bombers would not get through to drop their deadly bomb loads. Defence at night was by derivatives of the early jet fighters, which had even poorer performance due to the additional weight of the AI radar. Delays to the development of the second-generation Hunter and Swift single-engine day fighters, and the missile-armed Javelin all-weather fighter, meant that the USAF sent F-86 Sabre swept-wing fighters to Britain for both day and night fighting to fill the capability gap until the Hunter and Javelin entered service.

Meanwhile Britain's offensive capability was with World War 2-vintage technology Lincolns, replaced in the interim by US-supplied Washingtons, until the Canberra jet bombers could enter service. At least the Canberras had the altitude and speed to operate effectively.

While the RAF was building up its defending capability there was a high level of support from the USAF, as following the Sabres were F-84s and F-100 Super Sabres, some being capable of carrying tactical nuclear weapons. These were progressively replaced by F-101 Voodoos, multi-role F-4 Phantoms, F-111s and F-15s, the latter still in service. The USAF provided strategic bombers initially in the form of World War 2-vintage B-29s and the slightly improved B-50s. Then came the massive B-36s, followed by the jet-powered swept-wing B-47s which had a performance often superior to the defending fighters. As the Cold War came to a close the larger B-52 took over the strategic offensive, and continues in global operations in the conventional role.

Meanwhile the RAF fighter defences were progressively improved with Lightnings taking over from the Hunters, but with less range than the earlier Meteors. As a result in-flight refuelling became a regular routine to keep the interceptors aloft to intercept the hostile Soviet bombers and long-range reconnaissance aircraft. Following the Lightnings were the air-defence Phantoms, which were withdrawn when the air-defence version of the Tornado entered service. Both these types were very capable aircraft with long-range air-to-air missile armament and capable of taking on multiple hostile targets at a time. With the final retirement of the air-defence Tornados, the QRA continues using Typhoons, with little changed since the supposed end of the Cold War.

The RAF strategic bombing capability was centred on the V-Bomber fleet, with the Valiant as a short-term solution for early delivery, while the more capable Vulcan and Victor were in development. Both these types with their crews maintained Britain's nuclear deterrent with aircraft on stand-by ready to go at a moment's notice on what was almost certain to be a one-way trip. When the nuclear deterrent was taken over by the Royal Navy Polaris submarines, Victors became the tanker in the sky and the Vulcans took over the conventional bombing role, until their one and only action in bombing the runway in the Falklands following the Argentine invasion. The RAF strike capability transferred to smaller twin-engine aircraft such as the Buccaneers, followed by the Tornado GR.4s which are still in service, but with conventional weapons.

Left: Wymeswold still has the three runways, although they are obstructed in places. The gutted ATC building and associated structures, including the water tower and a T.2 hangar remain on the original technical site, and a very active business park covers the remainder of the old RAF camp. *Author*

Left: Air Traffic Control varied on Cold War airfields, but the standard World War 2 style continued in use, particularly on RAF Fighter Command stations. This example is still in use at Duxford, and many others still in existence have been preserved as heritage sites. *Author*

The much reduced RAF will soon have a very effective transport and flight refuelling capability, with increased numbers of helicopters, but reduced strike capability and no maritime reconnaissance role at all. This will result in further closures of former Cold War airfields, some of which may be taken over by the Army as it withdraws from Germany. The RAF will be less able in the future to participate in regional conflicts such as Afghanistan, although it is seen as vital to our nation's security. What once was a global force is now reduced to a scale comparable to many smaller countries, while Britain still claims to be a world power. When vital capabilities are lost, it takes many years and heavy budgets to recover them, if ever.

Although much of the desktop research for the text and captions comes from many reference sources, the picture research demands a wide search for material. I should like to thank Peter Elliot and Gordon Leith of the RAF Museum, Ian Proctor and his staff at the IWM, Kate Pyne of the AWE and Eric Snowdon for their help in sourcing illustrations. Also appreciated was the permission by Robert Pooley to use a selection of airfield charts in the current Pooley's Guide, and thanks are also due to both Margaret Pettitt and Andrea Stumpf for permission to use some vintage copies in my collection of Jeppesen Approach Charts. Michael Bowyer came to my rescue with illustrations of early USAF Cold War jets at airfields where I had no previous coverage, for which I am grateful. I should also like to thank Peter Green for his help with s of early Canberras and Lincolns and Andrew Thomas for his s of Washingtons. Many thanks go to my wife Martha who has been fully supportive throughout, including setting up the systems and buying me a splendid professional scanner to remove the risk of damage or loss to many priceless pictures taken over the years. I would also like to thank the many people who have allowed me over many years media access to RAF stations and USAF bases around Britain. Without their help the picture search would have been far more demanding and less fruitful. The biggest challenge has been to source enough illustrations of the RAF tactical bombers, such as Lincoln B.2s, Washingtons and early Canberra Wings. Although it will not help in the case of this book, the author would be keen to obtain pictures of these types as well as additional pictures of USAF B-29/50s, B-36s and B-47s at Fairford, Brize Norton, Greenham Common, Upper Heyford, Bruntingthorpe and Chelveston.

Philip Birtles,
Stevenage,
June 2012

Right: Meteor F.8 of 600 Squadron RAuxAF on finals to land at Biggin Hill. The squadron flew Meteor F.8s from November 1951 until disbandment on 10 March 1957. *RAF Museum P013526*

Chapter 1

The Berlin Airlift

Strictly speaking, the duration of the Cold War spanned the period from the Yalta Conference in February 1945 to the dissolution of the Soviet Union on Christmas Day 1991. In more practical terms it was probably started by the Soviet blockade of Berlin, which resulted in the Berlin Airlift between 24 June 1948 and 12 May 1949. Tensions were further increased by the closing of the borders between East and West Berlin and the start of construction of the Berlin Wall on 12/13 August 1961, while during the Cuban Missile Crisis a year later nuclear war appeared to be on the brink of starting. The Berlin Wall finally started to come down on 9 November 1989, and 11 months later East and West Germany were reunited as one nation.

The Cold War was different to any other previous potential conflict, as both the opposing superpowers had developed nuclear weapons. The Americans had tested their first atomic device in New Mexico on 16 July 1945, believing they were well ahead of the Soviet Union. The first practical application was by the USA against Japan, where two early developed atomic bombs were dropped, the vast destruction resulting in the end of World War 2. Then on 4 September 1949, an American Radiation Service B-29 over the Pacific detected a large radioactive cloud, which confirmed that Russia had tested an atomic bomb somewhere over the Asiatic mainland between 26 and 29 August.

Meanwhile in 1947 the British Government made the decision to develop its own nuclear weapons and the Air Staff under the leadership of Lord Tedder were responsible for drafting the specification for the weapon and the aircraft to carry it. Since 1946 the McMahon Act had forbidden America to share nuclear information with its allies, but the US Government gave active encouragement for Britain to develop her own independent nuclear deterrent. The first British atomic device was successfully tested on a Pacific island on 3 October 1952, followed by the design of a suitable casing and warhead during the winter of 1952/53. Following a series of dropping trials of inert weapons, the assembly of Britain's first nuclear weapon continued throughout the summer of 1953, reaching its operational form as the 24ft-long 10,000lb Blue Danube. This was successfully tested at Woomera in Australia on 14 October 1953, producing a yield of 40 kilotons, but it was another three years before an air drop test was carried out. The first components of Blue Danube were delivered to Wittering in November 1953, but the V-Bombers to carry it would not follow until mid-1955, when Blue Danube would be obsolete, leaving Bomber Command seriously lacking in a strike capability. Fortunately President Eisenhower led an exchange of nuclear information between America and Britain, and the resulting more refined American nuclear weapons were to be carried by the RAF V-Force for much of its existence.

The RAF heavy bomber at the start of the Cold War was the Avro Lincoln, which was a development of the World War 2 Lancaster and planned for use against Japan before the atomic bombs brought the conflict to an abrupt halt. Carrying a full bomb load of 14,000lb and full fuel, the speed was only 230mph, operational ceiling 22,000 feet and a range of 2,250 miles. The Canberra jet bomber would not be ready for service until 1951 and it was a light bomber with insufficient range to reach targets deep in Soviet territory. As an interim measure 87 USAF-surplus B-29s were supplied to the RAF, named Washington. The Washington had a range of 3,250 miles, a top speed of 357 mph and with its pressurised crew compartment could reach 30,000 feet. Its one drawback was that its maximum bomb load was only 12,000lb.

In 1944 four USAAF B-29s had made forced landings in Soviet territory due to having sustained battle damage while attacking targets in Japan. Although the crews were returned to the USA, the B-29s were not, and Stalin gave the task of reverse-engineering these aircraft to Tupolev, who produced the Tu-4 with the first three prototypes ready for flight testing in early 1947, powered by reverse-engineered engines. The Tu-4s were first shown publicly at the major Soviet air display near Moscow on 1 May 1948, by which time the first examples had already been delivered to the Soviet Air Force. This gave Stalin the capability of one-way attacks against the USA, but they still did not have an atomic bomb. By 1954 the Russians had caught up with bomber technology, as during the Tushino flypasts on 1 May two new large jet bombers were paraded. The first was a large swept-wing aircraft with four jet engines buried in the wing roots. NATO codenamed it 'Bison' with the Soviet designation Myasishchev M-4, and the other was a twin-jet swept-wing designed by Tupolev and codenamed 'Badger'. The following year a further new bomber was the long-range Tupolev Tu-95 'Bear' which had swept wings and was powered by four large turbo-prop engines. With these bombers the Soviet Air Force had the capability of delivering its increasing stocks of nuclear weapons, although it would take until the end of the 1950s to have sufficient numbers of the new bombers.

The first American nuclear weapons were delivered to USAFE bases in Britain in 1954 due to the known advances in the Soviet weapons build-up. The RAF stockpile was growing at a much slower rate, both governments concentrating on tactical rather then strategic weapons. By mid-1954 the component parts of the weapons were in a secure store at Sculthorpe where the 47th BW B-45s were based, at Woodbridge and Wethersfield with the 20th FBW F-84Gs in residence, as well as at Bentwaters and the major SAC

Left: With the blockade of Berlin by the Soviet Union of all surface transport to the city, the Allies conceived the Berlin Airlift to keep the population supplied with essential food and fuel. The airlift started on 24 June 1948 with the RAF initially carrying supplies by Dakotas. Dakota KN274 was being loaded with a spare Rolls-Royce Merlin engine at RAF Honington before crossing the Channel to Europe. *IWM R.2105*

bomber bases. By this time the weapons had been reduced in weight and an F-84G could carry externally a tactical nuclear weapon of 2,000lb, one-fifth the size of the bomb dropped on Hiroshima. To ensure there would not be an accidental nuclear explosion, the components of the weapons were stored in widely separated hardened dome-shaped bunkers. The same principle was used with the bombers, as on training flights the SAC aircraft were not flown with the weapons in the strike configuration, avoiding potentially catastrophic accidents in the event of a crash.

With the build-up of the superpowers' nuclear arsenal, any resulting World War Three would in effect have been the end of civilisation as we know it. With a first strike detected from one side, the opposition would launch a retaliatory attack against predetermined targets, resulting in mass destruction on both sides. Life in the Northern hemisphere would mainly cease to exist, all utilities would be disabled, and a few people may be alive for a period in nuclear shelters, but with little chance of long-term survival. The nuclear clouds would no doubt eventually drift into the southern hemisphere, making survival there almost impossible. There was a time early on when America believed itself to have enough strength to destroy Russia by nuclear attack without any retaliation, but fortunately sense prevailed and a nuclear holocaust was avoided. No other actual, or potential conflict, had the capability of destroying the entire human race.

Between 1954 and 1956 the European defence policy changed drastically, and by 1956 the launch of an all-out nuclear attack was unthinkable. The defence policy changed to a European shield of conventional forces with massive tactical air support. The USA, meanwhile, was conducting nuclear tests to produce weapons for tactical purposes, making the bombs with reduced nuclear fall-out, resulting in more of a military weapon rather then one of mass destruction.

In the early 1970s, with the doctrine of massive retaliation having lost credibility ten years previously, NATO adopted a strategy of flexible response, which consisted of a range of possible retaliations to Soviet aggression, ranging from conventional weapons, tactical nuclear weapons to full strategic nuclear retaliation. NATO now depended on this vulnerable triad for deterrence, although it gave NATO planners a flexible response to aggression, where any retaliation could be escalated as necessary. As the 1970s progressed it became clear to the Russian planners that it was unlikely that NATO would authorise the use of nuclear weapons due to the horrific consequences. The Soviets believed that NATO would only defend itself initially with conventional weapons, and that nuclear weapons would only be used in desperation to avoid defeat. To counter the threat of Soviet long range theatre nuclear weapons NATO filled the gap in 1977 with the 48th TFW F-111s at Lakenheath to complement the 72 F-111s at Upper Heyford and some 50 Vulcan B.2s with RAF Strike Command at Scampton and Waddington.

By the mid-1970s NATO intelligence had become aware that the Soviet Union was developing solid-fuel-powered ICBMs launched from hardened underground silos. The first of these was the SS-18 ICBM, the first deployment of 308 missiles in 1974 able to carry ten 500-kiloton MIRVs over a range of 7,500 miles and targeted on the American Minuteman ICBM installations. Following the SS-18 were two more highly accurate ICBM systems, the SS-17 and SS-19, both of which could deliver up to six 500-kiloton MIRVs over ranges of 6,200 miles, some of which were aimed at targets in Europe. In 1977 the SS-20 IRBM deployment upset the finely balanced East-West strategy, since where the larger systems were fired from fixed silos which could be targeted, the SS-20 was mobile and very difficult to locate. NATO had no counter to the SS-20 with its range of up to 4,600 miles, depending upon the weapons load, and therefore all of Europe was under threat from within the Soviet Union. The only counter to this weapon system was the F-111s based at Lakenheath and Upper Heyford, but these bases would be the first to be targeted by the SS-20s. What was therefore required was a NATO mobile missile system which could be widely dispersed and carry a nuclear warhead, restoring the credibility of the flexible response and freeing up the F-111s for conventional operations. The Soviet Union had also developed the Backfire and Fencer bombers carrying long-range stand-off nuclear weapons which would destroy NATO air bases losing the capability of theatre nuclear response. As a result at a NATO meeting on 12 December 1979 the decision was made to develop and deploy 464 ground-launched cruise missiles (GLCM) in a number of allied European countries including Britain. Each missile would be armed with a single nuclear warhead and in Britain they were allocated to Greenham Common and Molesworth from where they could be deployed in advance of a build-up of tension to predetermined sites throughout the southern part of the country.

The Yalta Conference

The Yalta Conference was the second of three conferences between the heads of state of America, Britain and the Soviet Union. The Conference, held between 4 and 11 February 1945, was convened in the Livadia Palace near Yalta in the Crimea and was attended by President Franklin D. Roosevelt, Prime Minister Winston Churchill and General Secretary Joseph Stalin, its main purpose being to discuss the re-establishment of the nations in Europe, and especially to create an agenda for the governing of post-war Germany. Churchill was highly suspicious of the motives of Stalin, but Roosevelt believed that if he gave everything he could without asking for anything in return, Stalin would work with the Allies towards a world of democracy and lasting peace. By the time of the conference, Soviet forces were through Poland and Hungary and within 41 miles/65 km of Berlin, resulting in Stalin being in such a strong position that he could dictate his own conditions.

Each leader had his own agenda for the conference, Roosevelt requesting Soviet support in the US war against Japan, including an invasion; Churchill specified free elections and democratic governments in occupied Europe, particularly Poland; and Stalin demanded a Soviet sphere of influence in eastern and central Europe as essential to the security of the USSR, including Poland. In return for assisting the USA, Stalin wanted American recognition for an independent Mongolia from China which was agreed, and the USSR would enter the Pacific War three months after the end of the war in Europe. Roosevelt agreed to Stalin's conditions, hoping that the United Nations (UN) could counter Soviet domination, but the overall impression was that the Yalta Conference allowed the USSR to maintain its control of Eastern Europe through to Asia, leading to the formation of the Soviet Bloc. The Soviets also agreed to join UN provided all permanent member countries had the power of veto, in effect blocking any unwanted decisions. By this time the Red Army occupied most of Eastern Europe with forces three times the size of the Allied forces in the west.

The Big Three ratified previous agreements for the post-war three zones of Germany, with the later addition of France in areas ceded by Britain and the USA. Although Berlin was within the Russian zone, it would be divided into three sectors shared by the Americans, British and Soviets. The immediate priority was the unconditional surrender of Nazi Germany, the country to be demilitarised and reparations made to invaded countries by the use of German forced labour. The status of Poland was discussed with agreement to reorganise the communist provisional government on a broader democratic basis, but Stalin never honoured his promise to allow free elections. It was agreed that democracies would be established in all liberated and former Axis satellite countries with free elections held and order restored. The declaration, however, contained no mechanisms to enforce the agreed principles.

The Soviet Union annexed the three Baltic countries already occupied by them as Soviet Socialist Republics while Poland, Hungary, Czechoslovakia and Albania all became satellite states, followed later by East Germany. In due course Britain and the USA recognised the communist-dominated regions, cancelling out the substance of the Yalta Declaration. With the death of Roosevelt, President Truman authorised the dropping of an atomic bomb on Hiroshima on 6 August 1945, two days before the USSR had agreed to support the USA in the war against Japan.

Left: The Yalta Conference was held in the Crimea between 4 and 11 February 1945 with the General Secretary of the Soviet Union, Joseph Stalin, seated on the left, US President Franklin D. Roosevelt across the table and British Prime Minister Winston Churchill, who is not in this picture. The aim of the conference was to define the government of postwar Germany.
IWM NAM 211

The Berlin Blockade

As the first major crisis of the Cold War, from 24 June 1948 the Soviet Union blockaded the Western Allies' road and rail access to the Berlin sectors under their control, the aim being for the Soviet zone to supply the city with food and fuel, giving the USSR practical control over the entire city. In response, the Allies organised the Berlin Airlift to carry over 4,000 tons of supplies per day by air using RAF, Commonwealth air forces and the newly formed United States Air Force (USAF) cargo aircraft. In the period between 24 June 1948 and 12 May 1949 over 200,000 flights were made, carrying 13 million tons of supplies.

With the end of the war in Europe on 8 May 1945, the Soviet and Western forces were facing each other along a line roughly following the River Elbe, with variations in several locations. Units of the reforming French forces were also present in south-west Germany. From 17 July to 2 August 1945 the Allied powers ratified the Potsdam Agreement, defining the constituents of post-war Europe. As previously, Germany was to be divided into four temporary occupation zones based around the current locations of the allied armies. Berlin was located 100 miles within the Soviet zone and was to be divided into four sectors, this time including the French. The administration of occupied Germany was to be coordinated by the Four Power Allied Control Council.

In the Soviet zone, the Communist Party and the Social Democrats were forcibly unified for the establishment of a democratic anti-fascist regime, with all other political activities suppressed. Key industries including armaments with equipment and skilled personnel moved to the Soviet Union. The area of the Soviet zone was able to provide much of Germany's food, while the British and American zones were always reliant on food imports. Stalin had incorporated a large part of eastern Poland into the Soviet Union, and provided some compensation by ceding to it a large area of Germany east of the Oder-Neisse line which had contained much of Germany's fertile land.

Stalin expected to undermine the British occupation of their zone, and that America would withdraw within two years, leaving no barriers to a unified Germany under Soviet communist domination. A further factor contributing to the blockade was that there never had been a formal agreement guaranteeing surface access to Berlin through the Soviet zone.

Above: Upon the death of Roosevelt, Truman became President of the USA. He attended the Potsdam Conference with Stalin (left) and Churchill (right) between 17 July and 2 August 1945. The aim of the conference was to define the political map of Europe following the end of World War 2, with Germany being split into four zones, and Berlin into four sectors, shared by Britain, the USA, France and the USSR. *IWM BU.8988*

Right: During the Berlin Airlift, RAF Dakotas used Tempelhof Airport in Berlin as the cargo terminus. The sloping floor of the aircraft due to its tailwheel configuration made unloading slower, but the larger York, with a more level floor allowed more rapid turnarounds. The USAF used nosewheel-equipped C-54s which had a door sill at truck bed height, making turnarounds very rapid. *IWM HU.36826*

Left: With their hulls protected against corrosion, the RAF Sunderlands were particularly suitable for transporting salt to the beleaguered city, landing on Lake Havel to reduce congestion at the other busy airfields. Sunderland MR.5 of 201 Squadron was being unloaded with vital supplies for the Berlin population. The blockade was cancelled by the Soviets on 12 May 1949, although the airlift continued to build up stocks of fuel and food in case the border was closed again.
IWM MH.30687

Allied leaders had mistakenly relied on Soviet goodwill to provide a right of access to the Berlin sectors. It was believed that the Soviet restriction of allowing ten trains a day along one rail route was just temporary, but they refused any further expansion requested by the Western Allies. In addition only three air corridors were permitted to be used by the Allies to fly to Berlin from Hamburg, Buckeburg and Frankfurt. In 1946 the Soviets stopped shipping agricultural materials to Berlin from their zone, and in reply the Americans did likewise with industrial units from western Germany to the Soviet Union.

Berlin became the focus for Western and Soviet efforts to re-align Europe to their respective visions. The city had suffered enormous damage during the war, with its population reduced from the pre-war 4.6 million to 2.8 million, and the Western Allies were not permitted access until two months after the surrender, the local population suffering badly from the Soviet occupation. As a result when local elections were held in mid-1946, there was a massive anti-communist protest vote in the Soviet sector of Berlin with an overwhelming election of democratic members to the city council.

My father was an Army Captain in the Royal Engineers as part of the Allied occupation of the British sector of Berlin, and I was fortunate to be able to visit him with my mother in the summer of 1947. We travelled by sea from Harwich to the Hook of Holland and then by train to Berlin. Along much of the train journey there was still the wreckage alongside the track of trains destroyed by the Allies, but when travelling through the Soviet zone the windows were shuttered to hide the activities outside. On arrival in Berlin, we stayed in my father's apartment and we could move around the local area. Two years after the end of the war many of the streets were still lined with massive heaps of rubble behind which were the shells of once fine buildings. The Reich Chancellery with Hitler's flooded bunker close by was a ruin, and the border with the Soviet sector was the Brandenburg Gate, guarded by Soviet soldiers. Because of a shortage of fuel, not only were trees cut down, but the roots were dug out of the sandy soil as additional fuel, leaving nothing to waste. Allied troops had a form of money, but the real currency consisted of cigarettes, coffee, chocolate and family heirlooms, making a very fragile economy. Having arrived from Britain which was still suffering from rationing, I was surprised on a visit in an army truck to the Erz Mountains, to sample my first cream cake!

The Western Allies were keen to rebuild the German economy to allow the country to become an independent nation as soon as possible, but this was strongly resisted by the USSR, fearing western political and economic penetration through aid. In late January 1948 the Soviets began to stop British and American trains to Berlin to check the identities of the passengers. Meanwhile the Western Allies announced on 7 March 1948 the intention to finalise the economic merger of the German western occupation zones and establish a federal government. On 9 March Stalin and his military advisers decided to force the Western Allies to comply with the policies of the Soviet government by regulating access to Berlin, and the Soviet delegation walked out of the Four-Power Control Council meeting on 20 March, leading to a major crisis for Berlin. On 25 March the Soviets announced the restriction of the movement of all traffic between the British, American and French zones into and out of Berlin and no cargo could be shipped out of Berlin by rail without Soviet permission, the restrictions to commence on 1 April. As a result what was known as the 'Little Lift' was started with all military goods to be transported by air.

Tension continued to grow with Soviet interruption of surface transport over the next two and a half months, and in addition Soviet military aircraft violated West Berlin airspace by harassing the flights along the Allied air corridors en route to Berlin. As a result on 5 April a Soviet Yak-3 fighter collided with a BEA Viking civilian airliner near Gatow Airport, killing all the occupants of both aircraft, increasing tension even further. On 9 April the Soviets demanded the removal of all navigation beacons which defined the safe corridors for aircraft operations into Berlin, and on 20 April restrictions were also placed on water transport through the Soviet zone.

On 18 June the Soviet border guards halted all surface transport into Berlin, delaying all freight shipments, while the USSR government began economic sanctions in Berlin and held a major military exercise just outside the city. On 24 June 1948, the Soviets stopped all surface transport between the western zones and Berlin, and the next day ceased to supply food to the civilian population in the western sectors of the city, as well as cutting off the electric supply from the Soviet-controlled power-stations in their zone. At that time there were stocks of food in West Berlin to last for about 35 days, and coal for 45 days, and in the post-war rundown of the Allies' military might, personnel and equipment had been drastically reduced.

Although there had been no formal agreement between the four major allies to use the surface routes into Berlin through the Soviet zone, there was a written agreement on 30 November 1945 for the use of air transport. It had been defined that three 20-mile-wide air corridors would provide uninterrupted access to Berlin. The Soviets could not claim that the transport aircraft were a military threat and the only way for an airlift to be stopped was for the Soviet Union to take aggressive action against unarmed aircraft, creating enormous tension which would lead to reprisals by the Allies. It was a major challenge to carry enough supplies to feed the population of Berlin, and the task was given to General Curtis LeMay, commander of USAFE, whose initial reaction was that they could carry anything required. When the American military consulted the RAF about a joint airlift, they found that there was already an airlift in support of British troops still in Berlin. Following the experience of the 'Little Lift', it was possible to calculate the resources needed to support the entire city, the total loads being 5,000 tons daily, of which some 1,500 tons were food and 3,500 tons fuel.

With the post war reductions of US forces in Europe there were only two USAFE squadrons of C-47 Skytrains which would be capable of carrying 300 tons daily, and although the RAF were better prepared, they were only capable of carrying 400 tons of supplies daily. The RAF was able to bring in more aircraft quickly, while additional air transport was sourced from the USA, France and Canada. The RAF fleet was soon increased to 150 C-47s, known as Dakotas in British service, plus 40 Yorks based on the Lancaster bomber, which could carry up to 10 tons, raising the RAF contribution to 750 tons very quickly. Meanwhile the US military had some 565 C-54 Skymasters, the military equivalent of the commercial Douglas DC-4, which were capable of operating into Berlin airports. Although the eventual outcome of the airlift was unknown, the people of Berlin gave it their full support even though they might have had to make some sacrifices. The British and Americans decided to commence a joint airlift without delay and they were joined by the Australians in September 1948. The USAFE Operation Vittles was launched on 26 June 1948 with the first RAF aircraft in Operation Plainfare departing on 28 June.

By 1 July the operations were becoming established with Rhein-Main becoming a USAFE C-54 hub and Wiesbaden operating a mix of C-47s and C-54s. The stream of aircraft flew northeast along the American air corridor to land at Tempelhof Airport. On their return, to avoid confliction with the inbound stream, they flew due west along the British corridor, and once clear of the Soviet zone, turned south back to base. The RAF operated a similar system, flying southeast from a number of airports in the Hamburg area, also returning for home along the central corridor. On 6 July Sunderland flying boats joined the airlift, operating from Finkenwerder near Hamburg and landing on the Havel River near Gatow, their corrosion-resistant hulls being particularly useful for delivering salt to Berlin. In addition to British and American personnel, there were also aircrews from Australia, Canada, New Zealand and South Africa.

With no previous experience of a major airlift, the Allies were faced with a logistical challenge. Initially it was thought that it would all be over in three weeks, but it soon became clear that the blockade was going to be for the long haul. It was not only the logistics of delivering the food and fuel in the correct amounts to the correct place at the required time, but with the number of flights, control in the air was critical. The USAFE developed a complex block system with three eight-hour shifts of a C-54 followed by a C-47 section flying to Berlin. The aircraft were scheduled to take off every four minutes, flying 1,000ft above the group in front, the lowest group being at 5,000ft and the stack was repeated five times in steps behind. By the second week the daily loads had reached 1,000 tons.

Problems were created with regular maintenance of the transports and sometimes aircraft were not utilised in the most effective way, while crew rest and utilisation was somewhat irregular. It became apparent on 22 July that a long-term airlift was necessary, and Maj Gen William H. Tunner was appointed to command the operation, having had experience with the wartime Hump airlift to China in 1944/45. He reorganised the entire airlift operation, forming a combined task force with both the USAF and RAF. The USAF Military Air Transport Service immediately provided eight C-54 squadrons totalling 72 aircraft to Rhein-Main and Wiesbaden in addition to the aircraft already in service, and two-thirds of all USAF C-54 aircrew were transferred to the airlift.

On 13 August, Tunner decided to inspect the Berlin operation first hand, but while he was on the way a C-54 crashed and burned on the end of the runway at Tempelhof due to very low cloud and heavy rain which affected the performance of the radars. A second C-54 immediately behind the crashed aircraft burst its tyres on landing while trying to avoid the wreckage, and a third C-54 ground-looped on the secondary runway, closing the entire airport. Fortunately no one was killed and Tunner ordered all the stacked aircraft to return to base. As a result of this incident Tunner instituted new rules to overcome the problem of all-weather operations. All flights would be under IFR (instrument flight rules) at all times, whatever the weather, and if any aircraft missed its landing slot first time, it would return to base. This reduced accidents and delays. It was also found that it took the same time to unload 3.5 tons from a C-47 as it did 10 tons from a C-54. This was largely due to the sloping floor of the tailwheel C-47 which made truck access difficult, while the tricycle undercarriage of the C-54 maintained a level floor allowing cargo to be rapidly unloaded onto vehicles at sill height. The C-47s were therefore withdrawn and replaced with larger aircraft.

Turnaround was reduced to 30 minutes by restricting the crews to not leaving their aircraft during unloading, and providing refreshments and flying documentation direct to the aircraft. To further maximise the aircraft utilisation, the steps of aircraft groups were reduced to three minutes and 500ft of separation, the stacks being from 4,000 to 6,000ft, reducing the time to climb to designated altitude. Block times were shortened to six hours to fit in another shift, allowing landings in Berlin at every minute of the day and night. Regular maintenance became a priority with 1,000- and 2,500-hour and inspections to ensure full reliability of the hard-worked fleet.

The shortage of manpower was solved by the Berlin population who replaced most of the military personnel on the unloading, in return for additional rations, and as experience grew, the unloading was being achieved in less than ten minutes with some loads. By the end of August, only two months after the start, there were 1,500 daily flights delivering over 4,500 tons of cargo, and when the C-47s were replaced by an eventual total of 225 C-54s, the deliveries increased to 5,000 tons per day.

The original estimates of up to 5,000 tons of food and fuel would be sufficient to supply Berlin only in the summer weather. While the food quantity remained the same at 1,500 tons a day, coal to heat the city would be an additional 6,000 tons per day. By this time the RAF had the four-engine Hastings transport entering service from November and to help support the additional aircraft ex Luftwaffe ground crews were recruited. The main restriction in increasing the flight schedules was the limit on runway capacity of the two main airports in Berlin, with two runways at Tempelhof and only one at Gatow. Both runways at Tempelhof were resurfaced and improved in turn between July and October 1948 to allow operations to continue without a break. Gatow was similarly improved by the British by improving the existing runway and adding a second one using concrete. Although the French were short of transport aircraft, they did agree to build a new larger airport in their sector alongside Lake Tegel, the construction of which was completed in 90 days. The heavy equipment for the civil engineering tasks had to be dismantled and flown in sections, not only helping to build the airfield, but also to prove nothing could be kept out of Berlin. To make all weather operations more practical a Ground Control Approach (GCA) radar was installed at Tempelhof, with another at RAF-controlled Fassberg. Despite the provision of these extra aids, it was not possible to control the weather, and in November and December there was a blanket of fog across Europe, many aircraft failing to find their destination. On one day, out of 42 aircraft departing for Berlin, only one was able to land, and at one point there was only a week's supply of coal remaining. With an improvement in the weather in January 1949 over 171,000 tons were flown in and in March close to 200,000 tons were delivered, with nearly 235,000 tons in April. On 21 April the point was reached where the airlift was able to deliver more supplies than had previously been transported by train, and the growing success became a humiliation to the Soviets. On 15 April the Russian news agency TASS announced a willingness for the Soviets to lift the blockade, and soon after, the four powers began negotiations which resulted in a settlement reached on western terms. On 4 May it was announced that the blockade would end in eight days' time, and it was finally lifted one minute past midnight on 12 May 1949.

Surface transport began to move into Berlin, but the airlift continued for a period to build up stocks in Berlin and create a reserve in case the blockade was reimposed, the airlift officially

Right: The East German border guards were ordered to shoot any would-be defectors across the Berlin Wall. The wall was finally opened on 9 November 1989, reunifying Germany and resulting in the eventual collapse of the Soviet Union. *IWM HU.99550*

ending on 30 September 1949. In total the USA delivered 1,783,573 tons, while the RAF flew in 541,937 tons. The RAAF flew in 7,968 tons of cargo and nearly 7,000 passengers, and at the height of the airlift there was an aircraft arriving every 30 seconds. There were a total of 101 killed during the operation, including 40 British and 31 Americans in 17 USAF and eight British aircraft accidents. Out of nearly 700 aircraft used on the airlift over 100 were flown by 25 British civilian operators on charter and co-ordinated by BEA.

The Berlin Wall

The next major indication of an attempt by the Soviets to dominate Eastern Europe was the progressive erection of the Berlin Wall, isolating West Berlin within the Soviet zone of East Germany. There was also a separate longer inner German frontier which separated East and West Germany overall, both these borders symbolising what became known as the Iron Curtain between Western Europe and the Soviet-dominated Eastern Bloc. Before the Wall was built, some 3.5 million East Germans, representing some 20% of the German Democratic Republic's population, had defected from the east to West Berlin, giving them access to other European countries.

Up to 1952 there was relative freedom of movement between East Germany and the Western-occupied zones, but at a meeting on 1 April 1952 between the leaders of East Germany and Stalin, concern was voiced that the freedom gave access to Western agents, and that movements of East Germans should be controlled by the issue of passes to visit the west. As a result the border between East and West Germany was built up, including fortifications. The borders between East and West Berlin sectors remained open, but traffic was controlled. West Berlin became the stepping stone for East Germans to escape to the west. In 1955 the Soviets passed responsibility for border control to the East German government, and permission was granted for East German residents to visit their relatives in the west. However, there were massive defections, resulting in the East German administration restricting virtually all travel across the border.

The main problem for the German Democratic Republic (GDR) was that the people leaving for a better life were often young and well educated, as well as many professional people and skilled workers, resulting in a perceived 'brain drain' and loss of skills. By 1960 only 61% of the GDR population were of working age, compared with over 70% before World War 2, the loss of skills damaging the political credibility and economic viability of the GDR. At midnight on Saturday 12 August 1961, the East German security forces began to close the border between East and West Berlin, which was completed by the Sunday morning. The roads were made impassable to all normal vehicles and barbed-wire barriers were erected along the 97-mile/156km perimeter round the three western sectors and the 27 miles/43km between East and West Berlin. The first concrete blocks were erected from 15 August in addition to other obstacles including minefields. The East German security forces had orders to shoot anyone attempting to defect.

To maintain Allied rights to surface entry to Berlin, from 20 August 1961 an American battalion was rotated every three months, each travelling along the autobahn. When complete the Berlin Wall was 87 miles/140km long and in June 1962 an additional parallel fence was erected about 110yds/100m inside East German territory, creating a no-man's land offering clear fields of fire for the armed border guards. Construction of the final version of the wall consisting of reinforced concrete commenced in 1975, taking around five years to finish. There were eight border crossings through the wall between East and West Berlin, mainly for western nationals and Allied personnel to visit East Berlin, with visitors from the East allowed through provided they held appropriate permits. The crossings were allocated to different nationalities, and probably the most well known was 'Checkpoint Charlie' for use by Allied personnel and foreigners. Other border crossings existed between West Berlin and East Germany to allow transit between West Germany and West Berlin, using four autobahns, in addition to four railway routes and by water along canals and rivers. While the wall was in place some 5,000 East Germans were successful in defecting, with a reported 136 killed in the attempt.

Below: The Soviet Union closed the borders around Berlin at midnight on 12/13 August 1961, and progressively built a wall barrier with a no-go zone inside to deter any escape. The illuminated Christmas tree in the western sector contrasts with the stark security lights with almost total darkness behind the security area in East Berlin. *IWM HU.99563*

The Berlin Wall began to crumble and the Iron Curtain to be pulled back in 1989. On 23 August Hungary dismantled its border defences with Austria allowing over 13,000 East German tourists to escape into Austria. When the Hungarian government attempted to stem the flow of refugees and return them to Budapest, they blockaded the West German embassy and refused to return to East Germany. The East German government stopped any further travel to Hungary, but would allow those there to return. A similar incident happened in Czechoslovakia, but this time the refugees were allowed to leave on condition that they used a train route which transited East Germany. This was followed by protest demonstrations all over East Germany with the contrasting demands of 'we want out' and 'we're staying here' resulting in a 'Peaceful Revolution' with large protest rallies into the winter of 1989. Meanwhile the refugees continued to escape through Czechoslovakia and on 9 November it was decided to allow direct crossings from East to West Germany, available to everyone. The initial problem was that the security authorities were not informed and when the population arrived at the border posts they were still closed. However, the volume of people was so great that the guards finally gave way, overwhelmed by superior numbers, opening the gates with no identity checks. Although 9 November was when the Wall was breached, it took some time to complete the demolition; the initial action from the population was to attack the wall with sledgehammers, taking sections as souvenirs, and in some cases making large new unofficial border crossings.

The East German government opened ten new border crossings in mid-November, the crowds cheering as bulldozers demolished sections of the Wall to reinstate the original road routes. New border crossings continued to be opened into 1990, including the significant Brandenburg Gate on 22 December 1989. From 23 December West Germans were granted visa-free access to East Germany; prior to that they had been restricted to applications for visas which could take a week or more to obtain. The Wall remained a border with controlled crossings until on 13 June the East German security began the final demolition of the Wall which continued until November 1991. Meanwhile, on 1 July 1990, East Germany adopted the West German currency, allowing reunification to be formalised on 3 October.

The Nuclear Deterrent

Although these activities were severe restrictions to public liberties, on their own they were not part of the Cold War. The real threat was from the perceived Soviet nuclear capability in addition to its role of dominance in Europe. The Allied policy was by a form of trip-wire reaction, so that when a hostile Soviet attack was detected, the dispersed RAF and USAF nuclear-armed bombers would scramble before the Soviet bombs arrived and head for pre-designated targets in the USSR. In effect this would have been a one-way trip, as some of the targets would have been beyond the range of the bombers to allow a return, and it had to be assumed that following a Soviet nuclear attack, there would be nothing to come back to anyway. The American Government's nuclear deterrent strategy was based on a massive retaliation against the Soviet Union if Russia started an attack.

The plan, codenamed Operation 'Trojan', called for SAC to be launched into full-scale action with a fleet of B-47 Stratojets which flew faster than most fighters and could penetrate deep into the Soviet Union from Alaska and Western Europe. To support this strategy the 7th Air Division was formed in May 1951 and allocated to many of the 3rd Air Division bases in Britain. The reconstruction of four former RAF bases at Brize Norton, Fairford, Greenham Common and Upper Heyford was carried out to make them suitable for the operation of B-47s by extending the runways and building concrete bunkers for nuclear-weapons storage. In the summer of 1953 SAC began temporary duty postings (TDY) of an entire wing of 45 B-47s supported by 20 KC-97 Stratotankers kept at readiness in Britain for periods of 90 days. At the end of each TDY, another wing relieved the one in Britain, usually at an alternative base. These deployments became a major logistical nightmare for SAC and Military Air Transport Service (MATS), transporting thousands of personnel and tons of supplies to and from the USA in the few days required in each rotation. In addition, the B-47s were beginning to suffer from the intensive use, and in 1958 the TDYs were replaced by Reflex deployments which consisted of 20 B-47s from four wings deployed in Britain for three weeks at a time, during which period they were on full alert for one week, armed with nuclear weapons for instant departure. The introduction of Reflex deployments was intended to support the strategic concept of getting into the air one-third of the SAC bomber fleet within 15 minutes. To practise the alert status a simulated alarm was sounded almost every day, when the crews rushed to their aircraft, started the engines and rapidly taxied to the runway, before then returning to the dispersals. Only occasionally were actual flying alert scrambles made. By the end of 1958 SAC had a strategic fleet of more than 3,200 aircraft including 1,396 B-47s and 488 B-52s supported by 745 KC-97 and 322 KC-135 tankers respectively. The 15-minute quick-reaction alert resulted in 1,000 aircraft and crews being in a state of full readiness around the clock.

On 8 August 1953 the Soviet Union announced that it had developed the hydrogen bomb, which was a nuclear weapon with many times greater destructive power than the original atom bombs. A successful test of this new weapon had been made in Siberia early in August, much to the surprise of America, which had been monitoring for nuclear tests, but missed this one despite regular atmospheric research flights by specially modified B-29s and B-50s. The first American H-bomb was not tested until 16 November 1953. What was more disturbing was the news that Russia was busy developing long-range bombers capable of carrying the new weapon. These were the Tu-16 Badger with a range of up to 7,800km, the Tu-95 Bear with a range of up to 17,500km and the M-4 Bison with a range of 12,000km. To the surprise of western observers both the Tu-16 and M-4 flew over Red Square in Moscow on the May Day Parade on 1 May 1954. The Tu-95 flew four months later and went immediately into production. The Soviets planned to produce 25 Bisons a month, which was double the production rate of the USAF B-52s and it was predicted that the Soviet strategic air force would be greater than that of the USA by the end of 1959. As a result B-52 production was drastically increased, eventually reaching a total fleet of 744 bombers. This allowed the USA to threaten a Massive Nuclear Response against any Soviet threat with the manned bombers backed up by long-range Atlas and Titan Intercontinental Ballistic Missiles (ICBMs). By this stage the nuclear weapons of both sides were roughly equal, a balance which removed the American Massive Response from working as a nuclear deterrent. If one side made a premeditated nuclear attack, the other side would retaliate, which would be devastating and lead to the complete destruction of the territory of both sides. The large stocks of tactical nuclear weapons in Europe, which were intended for a Limited Nuclear Response, would equally lead to a global catastrophe as the combined destructive power of the weapons stockpiled in Europe was thousands of times greater than the explosive power of all the bombs used in World War 2, including the two atomic bombs dropped on Japan. The stocks of strategic nuclear weapons in both the USA and Soviet Union were far more than was needed, resulting in the failure of the Nuclear retaliation doctrine. When President Kennedy was elected in November 1960, the change was made to a Flexible Response Strategy.

The U-2 Incident

There were a number of incidents which tended to warm up the Cold War and threaten world peace. The first was known as the U-2 incident when on 1 May 1960 an American U-2 spy aircraft was shot down over the Soviet Union during the term of Dwight D. Eisenhower as president. The base of this spy flight activity was Peshawar in Pakistan, giving close proximity to Soviet Central Asia and allowing the monitoring of missile test sites and other military installations. The use of the very high altitude U-2 spy plane was to gather vital intelligence before satellites were orbiting the globe. On 9 April a CIA U-2 intelligence-gathering aircraft crossed the southern national boundary of the Soviet Union and flew over four Soviet secret installations including a Tu-95 strategic bomber base and missile test ranges, taking s. It was detected by the Soviet defences and unsuccessful attempts were made to intercept the intruder with MiG-19 and Su-9 aircraft.

Above: The American Government believed the U-2 spy planes were flying at such a high altitude that they were invulnerable to attack. On 1 May 1960, Francis Gary Powers was flying a U-2 on an intelligence-gathering mission across the Soviet Union, when the aircraft was brought down by a ground-to-air missile. From this time high-level penetrations of hostile territory became too risky, and nuclear bombers began to operate at low altitude to avoid detection by enemy radar for as long as possible. *Author's collection*

Following the success of this mission a follow-up was planned and another U-2, fuel, support and pilots were ferried on 28 April from their base at Incirlik in Turkey to Peshawar, including Francis Gary Powers as the mission pilot. When the weather was suitable on 1 May, Powers took off from Peshawar with a task to graph ICBM sites around Sverdlovsk and Plesetsk, followed by a landing in Norway. All the Soviet defence units were on full military alert from Central Asia to the extreme north of the USSR as the U-2 flight was expected. All air unit commanders were ordered to intercept the intruder and to ram if no other method was effective. Due to the extreme operational altitude of the U-2 it was beyond the reach of the piloted interceptors, but it was within the range of some SAM sites, a successful interception being made over the Ural Region by the first of three SA-2 Guideline missiles. Powers managed to abandon the aircraft successfully and was captured soon after landing.

One of the last targets overflown by the U-2 was the Chelyabinsk-65 plutonium production facility, the graphy of which would have allowed the USA to determine the number of nuclear warheads that were being produced in the USSR. Four days after the incident, NASA announced that one of its aircraft was missing north of Turkey, and that the pilot had reported difficulties with his oxygen, resulting in him becoming unconscious while the autopilot was engaged. Following this,

Soviet Premier Khrushchev confirmed that an American spy plane had been shot down, but made no mention of the pilot. Assuming the pilot had died in the crash, the Eisenhower administration issued a release stating that the aircraft was on weather reconnaissance and had accidentally strayed into Soviet airspace. The US administration stated that there was no intention of violating Soviet airspace, but on 7 May Khrushchev sprang his trap by announcing that the pilot was alive and the aircraft, although damaged, was intact, together with the surveillance camera and incriminating film.

The timing of the incident was unfortunate as it was two weeks before a Four Power Summit Conference in Paris between Eisenhower, Khrushchev, Britain's Harold Macmillan and France's Charles de Gaulle. The conference collapsed when Khrushchev walked out on 16 May, when Eisenhower refused to apologise for the incident. This was to have much wider repercussions than just the deterioration of relationships between America and the Soviet Union. It was realised that the high-flying bomber was no longer safe from interception, especially as it could be easily tracked by radar. Therefore the strategic bombers were changed to the low-level role to approach enemy territory below the radar for as long as possible, and instead of being painted an overall reflective white to protect against radioactive blast, they were painted in a low-level camouflage. This not only reduced the endurance of the aircraft, but caused problems with fatigue of the structure which had not been designed to take the stress of low-level gusts, the first casualty being the RAF Vickers Valiant V-bomber with wing structure fatigue failures.

The Bay of Pigs

Another American incident was the CIA-inspired Bay of Pigs invasion by a force of Cuban exiles to overthrow the Fidel Castro-led communist Cuban government. The Americans feared Soviet expansion, and for a Latin American country to be allies with the USSR was regarded as totally unacceptable. The USA had a policy that the European powers should not interfere with countries in the western hemisphere. On 15 March 1960, US President Dwight D. Eisenhower approved the plan to train, equip and lead Cuban exiles in an amphibious invasion of Cuba and on 4 April 1961, President Kennedy approved the Bay of Pigs operation following consideration of other sites. Recruiting of the anti-Castro Cuban exiles in the Miami area began in April 1960, followed by training and assessment of the candidates. A force of over 30 A-26 Invaders were prepared for offensive operations, with parachute, amphibious, armour and infiltration training undertaken in the southern US states and Central America. For air transport C-54s and C-46s were provided and on 9 April 1961 the personnel and equipment moved from Guatemala to Nicaragua.

In early 1961 the Cuban forces were equipped with Soviet-supplied T-34 and IS-2 tanks, self-propelled anti-tank vehicles, howitzers and other armaments. The air force was equipped with A-26 strike bombers, T-33 jet trainers and Sea Fury fighters. With all the training going on for a year, it was no secret that an invasion was planned and efforts were made to arm the civilian population as a 200,000-strong guerrilla army in addition to the regular 25,000-strong army and some 9,000 armed security police. This was against a raiding force of 1,500 Cuban exiles with two CIA agents. Radio Moscow broadcast in English on 13 April predicting the invasion within a week; in fact it happened on 17 April, just four days later. Indications were that the Cuban population were loyal to Castro and it was unlikely there would be any defections. The Soviet Union also supplied experienced advisors with World War 2 experience.

At around 06.00hrs on 15 April 1961 the first raid was made on three of Cuba's airfields by the A-26 bombers in the markings of the Fuerza Aerea Revolucionaria (FAR), the air force of the Cuban government. They were armed with bombs, rockets and guns and were flown by exiled Cuban pilots and navigators, the task being to destroy as much as possible of the FAR aircraft on the ground. The results were mixed, with civil aircraft destroyed as well as some military types. At the UN the Cuban Foreign Minister accused America of aggression against Cuba, which was vigorously denied by the US Ambassador to the UN. On the same day the security police began to round up and detain thousands of suspected anti-revolutionaries.

The next day, the FAR managed to prepare a number of the T-33s, Sea Furies and surviving A-26s for action to allow defence in air-to-air combat and against surface targets such as ships and ground forces. Late on 16 April the invasion fleet of the Cuban Expeditionary Force rendezvoused about 40 miles/65km south of Cuba, escorted by a number of US Navy ships. There was also a US Navy task force off the Cayman Islands including the aircraft carrier USS Essex. The task force sailed north without USN escort, while during the night the CIA organised an unsuccessful mock diversionary landing in an attempt to lure the defenders away from the Bay of Pigs.

Immediately after midnight on 17 April two CIA landing craft carrying underwater demolition teams entered the Bay of Pigs, followed by four transports carrying about 1,400 ground troops and armour, landing troops on two of the designated beaches. At dawn around 06.30 hrs FAR Sea Furies, T-33s and an A-26 began to attack the ships still unloading troops and equipment, one ship being so badly damaged that it was beached on the shore, the surviving troops being unable to fight due to the loss of their weapons. A pair of invading A-26s sank a Cuban Navy escort ship, and with two other A-26s attacked Cuban ground troops while providing a diversion for the drop of 177 paratroops. Another drop of troops and heavy equipment failed as the equipment was lost in a swamp and throughout the day more forces with armour were brought in to reinforce the initial defenders, while at 09.30 hrs one of the supply ships was sunk by a rocket attack from Sea Furies and T-33s. By about mid-morning the two surviving freighters pulled back south into international waters pursued by FAR aircraft. From midday to dusk, the defending forces built up their strength and advanced on the invading forces. Losses were sustained by the FAL A-26s with three shot down by FAR T-33s and two more diverted. At 21.00 hrs an air strike by three A-26s on one of the airfields failed due to poor visibility.

On Day 2, 18 April, Cuban regular troops and militia made advances as the invading forces withdrew and in the late afternoon, FAL A-26s attacked a supply column of buses and military vehicles with bombs, napalm and rockets inflicting heavy casualties, although the column later re-formed and continued its advance. During the night of 18/19 April FAL transports delivered arms and equipment to reinforce the invaders, leaving before daybreak. On Day 3 a final air attack was made by the A-26s, two of which were shot down by the defending forces, killing the American contract crews. Combat air patrols were flown by USN A4D Skyhawks from USS *Essex*, with all identities removed, to support the invading forces, but without engaging in combat. Without air support and with shortage of ammunition and equipment, the ground forces retreated to the beaches while being pursued by Cuban army tanks, artillery and infantry. Late in the day, two USN destroyers moved close into the shore to evacuate the retreating soldiers from the beach, until forced to withdraw by the ground fire, and searches continued until 22 April, during which another couple of dozen survivors were recovered. Between April and October hundreds of suspected rebels were executed, and many of the surviving invaders were swapped for tractors and others executed. The failed Cuban invasion was a political embarrassment to the Kennedy administration, in effect strengthening the Cuban revolution.

The Cuban Missile Crisis

The Cuban Missile Crisis was a serious confrontation between the USA and USSR in October 1962, which was probably the closest the world powers came to nuclear warfare. Fidel Castro was concerned that following the failed Bay of Pigs invasion, the USA might try to invade directly, and requested help from the USSR. The situation was not helped by the USA imposing an economic embargo on Cuba from February 1962. In September the Cuban government believed an invasion was imminent when the US government passed a resolution which authorised the use of military force in Cuba if there was a threat to American interests, followed by an announcement of a major military exercise in the Caribbean in October. Khrushchev agreed to support Castro by secretly placing Soviet strategic nuclear missiles in Cuba, as he did not want to lose the communist state in Latin America. To the Soviets, Kennedy was weak and indecisive and was expected to accept the missiles as a fait accompli. Kennedy's response was to warn that any nuclear missile launched from Cuba against any nation in the western hemisphere, would bring full retaliation from the USA against the Soviet Union. In addition there would be a strict quarantine of all military equipment, and any ships from whatever nations would be turned back if offensive cargoes were found.

The first medium-range ballistic missiles (MRBM) were delivered during the night of 8 September with a second batch on 16 September. The plan was to construct nine missile sites to include a total of 40 launchers with SS-4s and SS-5s, the latter with a range of 2,400 miles/4,000 km. Despite many reports from the Cuban people, they were not taken seriously by the US intelligence organisations, until an overflight by a U-2 on 14 October confirmed the construction of an SS-4 site. When Kennedy was shown the graphs of the site he called a meeting of the Executive Committee of the National Security Council (EXCOMM) to determine what action should be taken. The five possible courses of action discussed were: to do nothing; use diplomatic pressure on the Soviet Union to remove the missiles; an air attack on the missiles; a full military invasion; or more selective naval quarantine. There was unanimous agreement that a full-scale invasion was the only answer. However, Kennedy was concerned that the Soviets would retaliate with action in Berlin. There were suggestions that the missiles had an effect upon the strategic balance, but Secretary of Defence Robert McNamara pointed out that while the USA had some 5,000 strategic warheads, the Soviets had only 300. Therefore another 40 would not be significant in the military balance, but their close proximity certainly affected the political balance. A full-scale invasion

Below: No 19 Squadron operated Hunter F.6s from Leconfield from 29 June 1959, and converted to Lightning F.2s in December 1962. *MOD*

was not therefore an option, but a blockade was suggested as an alternative, although it was realised that this was an act of war. By 19 October four operational missile sites had been identified by U-2 flights over Cuba, and the US forces were put on to a high standard of alert, including Strategic Air Command (SAC) B-47 and B-52 bombers.

The restrictions to shipping fell short of a blockade, but remained as a quarantine action against all military equipment, with Latin American participation including support from Argentina, Venezuela and the Dominican Republic. On 22 October President Kennedy delivered an address to the Nation, revealing the discovery of the missiles in Cuba, while at the same time putting the US Forces up to DEFCON 3 level, which was reduced to DEFCON 4 on 20 November. On 23 October a proposal was made diplomatically to the US Ambassador in Turkey and to the US Ambassador to NATO, to consider the withdrawal of Jupiter IRBMs from Turkey and Italy in exchange for the Soviet missiles' withdrawal from Cuba. As the crisis deepened, on the night of 23 October, SAC was brought to DEFCON 2 status for the first time ever. This message was deliberately transmitted in the open to alert Soviet intelligence, and three radar systems were set up with direct contact to NORAD control, to detect any missile launches from Cuba.

Meanwhile Khrushchev had cabled Kennedy explaining that the Soviet Union rejected the arbitrary demands of the USA. He considered the blockade as an act of aggression, and the ships would be instructed to ignore it. Kennedy's response on 25 October stated that the USA was forced into action following repeated assurances that no offensive missiles were being deployed in Cuba. Work was reportedly continuing on the preparation of the missiles, as a result of which Kennedy ordered the loading of nuclear weapons on to the SACEUR bombers for a possible strike against the Soviet Union. The next day Kennedy was persuaded to delay the invasion option, and give more time for diplomatic and military pressures to have an effect.

Secret negotiations were going on between Aleksandr Fomin, the code name for the Washington-based Soviet KGB station chief, and John Scali of ABC News. Fomin asked Scali to communicate with his high-level friends in the State Department to see if the USA would accept a diplomatic solution to the crisis. Fomin suggested that if the Soviet Union gave assurances that the missiles would be removed under UN supervision, Castro would state publicly that he would not accept such weapons in the future, and in exchange the USA would publicly confirm that it would never invade Cuba. Brazil was asked to communicate with Castro that the US would be 'unlikely to invade' if the missiles were removed. Following the dialogue between Fomin and Scali, a message was received from Khrushchev along the same lines as the earlier discussions. Castro was convinced that an invasion was imminent and asked Khrushchev to make a pre-emptive strike on the USA, and all the Cuban-based anti-aircraft weapons were ordered to fire on any US aircraft. The CIA delivered a report that five of the missile sites appeared operational and the Cuban forces were mobilised for action.

The next day Khrushchev broadcast a message, contradicting his previous proposal, which suggested that he would authorise the removal of the Cuban missiles provided that the USA withdrew their Jupiters from Turkey and Italy. He further clarified this by stating that the USA was objecting to MRBMs on their doorstep, while they had installed IRBMs close to the Soviet borders. On the morning of 16 October a U-2 was shot down over Cuba by an SA-2 Guideline ground-to-air missile, killing the pilot. The U-2 had been sent over to test the Cuban defences, as the Cubans apparently did not have the capability to shoot down the aircraft, and if fired upon it would indicate that the Soviets had supplied suitable ground-to-air missiles. The successful interception of the U-2 was an indication that the Soviets were escalating the conflict and the US government were inclined to retaliate, and instead of discussions were prepared to attack. Fortunately they changed their minds in case the loss of the U-2 had been an accident, and Khrushchev had also reasoned that if a U-2 was shot down it would escalate the situation, and ordered the missile batteries not to engage the aircraft. It was learned that the decision to fire the missile was made by a Soviet commander acting on his own authority.

As the discussions got under way, Kennedy was persuaded that it was better to ignore the latest message and instead return to Khrushchev's initial offer. No part of the proposed agreement mentioned specifically the removal of the missiles from Turkey, but there was an understanding that they would be removed voluntarily after a settlement had been reached. In return for the Soviet missiles being removed from Cuba, the US Government would cease the blockade and give assurances not to invade Cuba in the future. The USA had little confidence that the deal would be accepted, and prepared for air strikes on the missile sites and economic targets, while also considering a Soviet reaction in Europe. After much deliberation, it was agreed secretly by Kennedy that the American missiles would be removed from Turkey and Italy in exchange for the Soviet missiles being removed from Cuba. On 28 October Khrushchev broadcast a message on Radio Moscow stating that the Soviet missiles would be withdrawn from Cuba, which was responded to by Kennedy stating that the USA would recognise the sovereignty of Cuba, and would not permit any invasion from US territory or other countries neighbouring Cuba. Because the withdrawal of the American missiles from Turkey and Italy was not made public at the time, Khrushchev appeared to have lost the conflict, and because Kennedy had won the contest between the superpowers, the Soviets had been humiliated.

France leaves NATO

On 19 December 1965 Charles de Gaulle was re-elected as President of France by a narrow majority. His policy was to make France a world power with a nuclear capability and therefore he had no further need for an American presence in his country. At a press conference on 21 February 1966 he announced that the ties with NATO would be reduced, followed two weeks later by the announcement that France would withdraw from NATO altogether. French forces were no longer available to NATO and would be withdrawn from Germany, while NATO HQ and all NATO troops must be out of France by 1 April 1967. As a result Operation "Freloc" (French Relocation of Assets) was instigated, which meant moving the USAFE HQ from Paris and the clearance of nine air bases and 77 other military installations, including radar systems, weapons stores and other assets, together with the relocation of 33,000 USAFE personnel and families. Amongst the units to move to Britain were the 66th TRW with three RF-101 Voodoo squadrons to Upper Heyford and three Fighter Wings also allocated to the 3rd AF in Britain. These were the 20th TFW to Wethersfield, the 48th TFW to Lakenheath and the 81st TFW to Bentwaters/ Woodbridge with a total of 225 F-100 Super Sabres and F-4 Phantoms. In addition, also based in Britain with the USAFE were two TRWs, the 10th at Alconbury and the 66th at Upper Heyford with some 100 RF-101s and RF-4s. Other USAF French-based assets were deployed to Germany and Spain, while some units returned to the USA, but still under the control of the USAFE with a high state of readiness.

Korean Air Lines Flight 007

A number of American aircraft were shot down when flying over the Soviet Union, but they were on obvious intelligence-gathering missions and had military personnel on board. However, one incident which escalated tensions between the USA and Soviet Russia was the shooting down on 1 September 1983 of a Korean Air Lines Boeing 747 operating flight 007 from New York to Seoul in South Korea, with its last stop at Anchorage in Alaska. All 246 passengers and 23 crew members were killed when the airliner strayed into prohibited Soviet airspace due to a navigational error and was shot down by Soviet jet fighters over the Sea of Japan, coming down near Moneron Island.

The Soviet Union initially denied shooting the airliner down, but later admitted to the incident, claiming the aircraft was on a spy mission, and accusing the USA of attempting to test the preparedness of the Soviet defences. The USA accused the Soviet Union of obstructing the search and rescue operations, and also of hampering the investigations by not releasing the flight data recorders until after a delay of nine years. As a result of the incident, tracking procedures for aircraft flying to Asia from Alaska were altered and the interface of the autopilots used on the airliners was redesigned to avoid a similar undetected error creeping into the flight programme. President Ronald Reagan ordered that the US military Global Positioning Systems (GPS) were to be made available to commercial aviation to ensure that similar navigation errors would not be repeated in the future.

Exercise Able Archer 83

Able Archer was a ten-day NATO exercise starting on 2 November 1983 across Western Europe which tested an escalation of conflict, resulting in a simulated nuclear attack. New unique coded communications were introduced with participation by heads of states, and a simulated DEFCON 1 nuclear alert was raised. With the deteriorating relations between the USA and Soviet Union, and the realistic nature of the exercise, it appeared to be a preparation for a genuine nuclear first strike. As a result the Soviets alerted their nuclear forces with the air force units in East Germany and Poland at readiness. The threat of nuclear conflict ended with the conclusion of the exercise on 11 November.

The Four Minute Warning

One of the features of the Cold War in Britain was the public alert system known as the "Four Minute Warning" which was in operation from 1953 until 1992, when the Cold War ended. Four minutes was believed to be the approximate length of time from when a Soviet nuclear attack could be detected until the impact of the weapons on their targets. This would result in the population being warned to take cover immediately by air-raid sirens and radio and TV broadcasts, although the warning of an attack would more likely be three minutes or less. Initially Jodrell Bank was used to detect and track incoming missiles, but from the early 1960s, the RAF BMEWS powerful radar installation at Fylingdales in North Yorkshire could track inbound missiles, allowing confirmation of probable targets. The responsibility for alerting the nation to an imminent air attack was with the UK Warning and Monitoring Organisation (UKWMO) at the UK Regional Air Operations Centre (UK RAOC) at the RAF Strike Command Operations Centre at High Wycombe. Linked into the warning system were 25 Royal Observer Corps (ROC) group controls, and in the event of radioactive fallout being experienced, warnings could be generated on a localised basis. In addition to ensure unbroken communications, the telephone links were hardened against EMP disruption.

The Cold War was therefore punctuated by a series of incidents of varying concern which kept the Western Allies and Soviet Union in a constant state of readiness, costing enormous amounts of money to maintain the arms race with the development and deployment of ever more deadly weapon systems. The Soviet Union did not have the resources to build vast fleets of strategic bombers, concentrating instead in defence against nuclear attack by deploying missiles to destroy any attack from the West. There was of course also the threat of Soviet nuclear submarines armed with atomic missiles, which were able to roam the oceans, creating tension worldwide. It was only by visiting Russia and other countries formerly within the Soviet Union after the break-up of the empire, that it could be seen that the Soviet threat hardly existed due to its poor economy. The threat had been a massive propaganda programme, and where construction sites had been established, the cranes were disused and rusting, having not been used for years.

Chapter 2

The Airfields

The Cold War, as far as the Western Allies were concerned was initially a tactical operation, followed by a strategic capability, when the more sophisticated aircraft and weapons had been developed. Immediately after the end of World War 2 the RAF was equipped with early Vampire and Meteor jet fighters, which in technological terms were not much advanced from the earlier piston-engine fighters. Speeds were certainly improving, but endurance was generally limited, requiring the development of flight refuelling to give greater ranges. These early first-generation jets were replaced by Hunters for day fighter operations and Javelins for night fighting, but although the speeds, armament and technology had improved, there was still a shortage of endurance. The Soviets would test UK defences by sending over high-flying bombers and reconnaissance aircraft off the coast of Britain, but with a service ceiling of between 42,000 and 45,000ft, there was little chance of the early jets catching these aircraft until the arrival in service of the Lightning.

RAF Bomber Command was initially equipped with Lincolns, a development of the famous Lancaster, and was in need of an interim strategic bomber capable of carrying a nuclear weapon while Britain's strategic V-Bombers were in development. This gap was temporarily filled by USAF-loaned B-29s, known in RAF service as Washingtons, while the

Above: Some of the Fighter Command stations still had Warren Truss Girder hangars with brick walls, wooden roof structure and wooden concertina doors. This is an example of the two-bay version with disused aerodrome beacon above the centre at Duxford. *Author*

Canberra tactical jet bomber was introduced to RAF service from May 1951, eventually undertaking a variety of additional roles including reconnaissance.

The RAF airfields therefore did not change greatly from their use during World War 2, the airfields selected being mostly the pre-war permanent bases with adequate accommodation for aircraft and personnel. Runways needed extending to cope with the longer take-off and landing requirements of jet aircraft, since they did not have the benefit of the lift generated over the wings by propellers, and more hardstandings were required, since the heavier aircraft coming into service were unable to operate from grass surfaces.

Left: In the mid-1920s the standard aircraft hangar was the 'A' Type with brick and steel structure and four doors on tracks which were opened and closed using a large winding handle. This example still exists at Upper Heyford, which was built in 1925. *Author*

Right: As a temporary protection for smaller combat types of aircraft the Blister hangar was developed during World War 2 and consisted of a steel curved frame with corrugated iron cladding. While some survive on Cold War airfields, they would only have been used for storage of equipment. The Blister hangar usually had a brick end structure and could come in any length required. This example is at Enstone, but is not in its original position. *Author*

Provision had to be made for the servicing and operation of jet aircraft, including provision of Avtur, or turbine fuel as an alternative to Avgas – petrol which was used in piston-engine aircraft, a number of which continued in service for many years, particularly in transport duties. Some of the less remote operational airfields were equipped with de-tuners, or sound suppressors, to deaden the sound of engines run on the ground. Security was taken seriously, with graphy not encouraged, including at early RAF air shows and Battle of Britain displays, although this was soon relaxed.

With the entry into service of the Mach 2 Lightnings which could intercept the high-flying Soviet bombers, the airfields where they were based had Quick Reaction Alert (QRA) hangars built close to the main runway threshold, where a pair of aircraft were maintained on constant readiness and scrambled when a potentially hostile Soviet aircraft was detected approaching Britain. Later the fighters were protected by Hardened Aircraft Shelters (HAS) which were dispersed around the airfield allowing routine maintenance and the aircraft to be at readiness, when armoured doors could be rolled back to allow the Phantom or Tornado to rapidly taxi to the runway for take-off.

With the arrival into service of the Valiant, Vulcan and Victor strategic V-Bombers, new engineering facilities were required and Operational Readiness Platforms (ORPs) were constructed on the side of the thresholds of the main runways, not just at the bomber bases, but also at other major airfields where the V-Bombers would be dispersed in times of tension. These ORPs were sections of concrete alongside the runways where up to four V-Bombers could be kept at constant readiness with the crews close by, and scrambled within four minutes to make a retaliatory attack against an identified pre-emptive Soviet aggression. This would in fact mean that by definition there would not be expected to be a base to return to, as they would have been designated targets by the Soviet bombers, as well as centres of industry and population.

Meanwhile the United States Air Force Europe (USAFE) occupied a number of bases in Britain which were extensively improved to handle the American strategic bombers then in development, as well as the new swept-wing jet fighters starting with the F-86 Sabre. Although this book is concentrating only on Britain's Cold War airfields, operations by the RAF and

Right: An easily erected World War 2 hangar was the T.2 which was a metal structure with metal cladding and six doors at each end running on tracks. Many still exist and complement existing hangars on Cold War airfields. They can be erected to whatever length is required, and can be found used for storage on a number of disused airfields. This example is at North Weald. *Author*

Left: A typical Expansion Period guardroom just inside the main gate at Honington. *Author*

USAFE were very much integrated within Europe particularly with the establishment of the North Atlantic Treaty Organisation (NATO) which was signed in April 1949. The initial membership was Britain and the USA, joined by Belgium, Canada, Denmark, France, Iceland, Italy, Luxembourg, the Netherlands, Norway, Portugal and Spain. Greece and Turkey joined NATO in 1952, with West Germany admitted in 1955. Aggression against any member was considered to be against the whole, resulting in a co-ordinated military response from the entire membership. France was later withdrawn by General de Gaulle, but is now in the process of rejoining. The opposition was not just the Soviet Union, but the communist-regime Warsaw Pact nations throughout Eastern Europe dominated by the Soviets.

USAFE fighter strength was improved, initially with F-100 Super Sabres and the all-weather F-101 Voodoos, later being replaced by Phantoms. The American Strategic Air Command (SAC) rotated their nuclear bombers to British airfields, the initial examples being the massive propeller-driven B-36 Peacemaker, of which some 400 were in service from 1948 until 1959. America's first strategic jet bomber was the Boeing B-47 Stratojet, of which 399 were built, to be followed by the same manufacturer's B-52 Stratofortress, late models of which are still in service today. In addition the General Dynamics F-111s provided a strike and reconnaissance capability as well as U-2s/TR-1s and Blackbirds for strategic reconnaissance. The USAFE bases were self-contained communities, accommodating American service personnel and their families. Included was the PX and commissary where they could buy any goods and supplies tax-free and pay with dollars. They had their own sports and entertainment facilities, so that there was really no need to integrate with the local population if they did not want to. However, the base commanders would encourage communications with local people to encourage good relations, and this was assisted considerably by many of the bases having an annual open house and air show when it was possible to sample some American culture at first hand. The USAFE bases were therefore rather more developed than their RAF equivalents, with vast areas of concrete and a number of specialised buildings located within the original site. Some were changed out of all recognition from their original layout, and while many have now been closed to flying operations, others are either used by the RAF or continue in USAFE service.

Left: The station HQ was usually located close to the main entrance at the Expansion Period RAF stations, and this brick-built version faces the main gate at Honington. The wartime camouflage can be seen on the original brickwork, but is not on the first-floor additions at each end. *Author*

Left: Temporary accommodation in World War 2 and after was in Nissen huts with brick end walls and corrugated iron curved roof. This example is one of a number surviving at Shepherd's Grove on the technical site which also includes a T.2 hangar, most of the surviving buildings being converted to commercial use. *Author*

Right: A slightly improved administration accommodation on some RAF World War 2 stations was the sectioned concrete structure buildings, again intended to be temporary, but which not only were used on many Cold War airfields, but still survive today in regular use. This example is one of a number in use at RAF Weston-on-the-Green. *Author*

Left: A typical Cold War ATC was rather more substantial, with the watchtower on the top giving a full view of the airfield and its approaches. This fenced-off example remains at Upper Heyford, the site being under the protection of English Heritage as a Cold War Heritage Site. *Author*

26   **UK Airfields of the Cold War**

Chapter 3

Tactical RAF Fighter Stations

Acklington, Northumberland;
NU230010, 3 miles NE of Felton by B6345

Acklington was a major fighter base of the northeast of Britain in World War 2 and was built on the site of a World War 1 airfield known as Southfields. With the start of World War 2 the airfield was transferred to Fighter Command as a satellite for Usworth. The station was used by a number of fighter squadrons resting and retraining from the rigours of the Battle of Britain in the south, and it was from here as well as other northern fighter bases that an attack by some 100 He.111s escorted by 40 Bf.110s was defeated on 16 August 1940, when attempting a daylight attack on Tyneside. Some 30 enemy aircraft were shot down and the raiders turned back without any damage. The facilities at the station were very basic, with a modest technical site and eventually three runways.

At the end of World War 2 the station was designated as a forward airfield in the Newcastle sector with Mustangs of 19 Squadron and Mosquito night fighters of 219 Squadron. The first jet fighters to arrive were Meteor F.3s with 263 Squadron which re-formed on 19 August 1945, but left for Church Fenton the following month. Acklington became a front-line fighter station again in 1957 with 29 Squadron equipped with Meteor night fighters and then Javelins until July 1958, and 66 Squadron Hunters from February 1957 until September 1960. The station was then transferred to the training role with 6 FTS Provosts and Jet Provosts until the FTS was closed in June 1968. Flying ceased in 1972 and the site was closed, the technical site becoming a prison and the airfield used for open-cast mining.

Right: Meteor F.4s were operated by 263 Squadron from Acklington from 8 August to 1 September 1949, after which the squadron returned to Horsham St Faith to re-equip with Meteor F.8s in October 1950. *Charles E. Brown*

Below: No 29 Squadron operated Meteor NF.11s from Acklington when it moved from Tangmere on 14 January 1957, where it re-equipped with Javelin FAW.6s from November the same year. *Air Ministry*

Biggin Hill, Kent;
TQ415605 5.5 miles SSE of Bromley

Famous as one of the prime Battle of Britain sector stations, Biggin Hill, located on a plateau on the North Downs, was first opened on 14 February 1917 when it became the Wireless Experimental Establishment in December responsible for the development of ground-to-air radio communications. With its location on the approaches to London, 141 Squadron Bristol Fighters were based there from February 1918 for the defence of the capital from hostile German Gotha bombers. In the inter-war period, the station continued in the fighter role until the autumn of 1927 when work started on significant improvements to the base. The original accommodation had been on what was to become the South Camp, with a completely new technical site being constructed on what was to

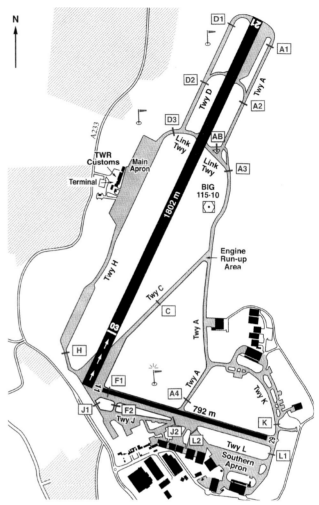

Right: Having opened as early as February 1917, the layout of Biggin Hill was certainly not traditional. The original three runways came together in the SW corner with the original technical site being along the southern boundary. In September 1932 the North Camp opened with Belfast hangars, which amongst many other buildings were destroyed by enemy action during the Battle of Britain. These were replaced by a pair of T.2 hangars in the North Camp, which was in effect to the west of the lower section of runway 03. Meteor jet fighters of 600, 615 and 41 Squadrons operated at Biggin Hill from 1950 and 41 Squadron re-equipped with Hunters in 1957, but was disbanded in January 1958. Biggin Hill closed as an RAF station in 1979, and is now a busy general aviation airport. The site of the RAF North Camp has been largely cleared and landscaped for commercial purposes with one T.2 hangar remaining, while the flying clubs and aircraft maintenance operated from the southern area. *Pooley's Flight Guides*

A pair of 41 Squadron Hunter F.5s on finals to land at Biggin Hill. The squadron flew Hunter F.5s from June 1955 until it was disbanded on 31 January 1958.
RAF Museum P013525

become North Camp. The new camp included Belfast hangars, workshops, stores, messes, barrack blocks and administration offices and was ready to receive aircraft when 23 and 32 Squadrons arrived with Bulldogs in September 1932.

Biggin Hill prepared for war with the arrival of Hurricanes for 32 Squadron in October 1938 followed by 79 Squadron in December. In May 1939 3 Squadron arrived with Hurricanes to complete the Biggin Hill Wing. It was one of the most damaged airfields during the Battle of Britain with the original Belfast hangars being destroyed and later replaced by a number of T.2 hangars in both the North and South Camps, but it was always operational. Three runways were also constructed in a rather unusual fan configuration, meeting at the south end of the airfield, the middle one eventually becoming disused. In early 1941 the Biggin squadrons switched to the offensive, but where the Battle of Britain had failed to close the station to flying, the arrival of the first V.1 flying bomb a week after D-Day put the station in the line of fire for London, and it fell within the perimeter of the London balloon barrage. The fighters returned in October 1944 with the Spitfires carrying out bomber escort duties on daylight raids on Germany. In June 1945 Biggin Hill was transferred from 11 Group Fighter Command to 46 Group Transport Command, the resident aircraft being Dakotas.

Biggin Hill returned to the day fighter role on 10 May 1946 when 600 (City of London) and 615 (County of Surrey) Royal Auxiliary Air Force (RAuxAF) Squadrons re-formed, the station being administered by Reserve Command. It reverted to Fighter Command in November 1949, the two RAuxAF units being re-equipped with Meteors during 1950 and were joined by a regular Meteor 8 equipped 41 Squadron on 29 March 1951, which had moved from Church Fenton.

In 1957 41 Squadron converted to Hunters, with the main runway extended to 2,005 yds/1,808 m but with the disbandment of the RAuxAF in March 1957, the station lost its strategic value and 41 Squadron also disbanded on 31 January 1958. The station which had been so key to winning the Battle of Britain therefore played a minor part in the Cold War. In late 1979 the decision was made to close the RAF station, and starting with the South Camp, civil aviation moved in from the closing Croydon Airport, and today Biggin Hill is a busy general aviation airport.

Church Fenton, Yorkshire;
SE530380 NW of Selby by B1223 road.

The station was constructed as a fighter base in the pre-war RAF Expansion Scheme, originally opening as a grass airfield in June 1937. Two 'C' Type hangars were built on the western side of the landing ground with enough room between them for a third to be added at a later date. This third hangar eventually became one of the lower cost T.2s. The remainder of the technical site and accommodation was constructed behind these hangars on both sides of an adjacent lane, the work finally being completed in 1939. The station had originally been allocated to 12 Group Fighter Command, but with the outbreak of World War 2 in September 1939, it became a sector station in 13 Group. The station was allocated the task of defending the East Coast of Yorkshire as well as the major towns in the West Riding. In addition to local defence, the station was also used by squadrons taking rest tours from

No 19 Squadron was based at Church Fenton from 23 April 1947 until moving to Leconfield on 29 June 1959. During its time at Church Fenton the squadron operated Hunter F.6s from October 1956 and XG172 B:19 Squadron was one of their aircraft. *J. J. Halley*

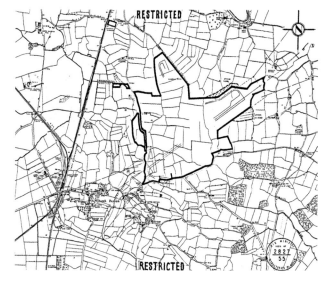

Left: In 1955 plans were made to extend the runways at Church Fenton to allow operations of the early jet fighters then planned for service with the RAF. The major extension was to the main runway to the northeast and a small extension to the north of the cross runway. *RAF Museum*

the rigours of the Battle of Britain, rebuilding their strength and training replacement pilots.

With the end of World War 2, 263 Squadron, one of the first to be equipped with Meteors, arrived from Acklington, and Church Fenton was retained as a permanent RAF station with the two concrete runways that had been built for fighter operations extended considerably to accommodate the early jet fighters. A number of other Meteor squadrons shared the base with night-fighter Mosquitos. On 18 October 1950, the Meteors of 609 (West Riding) Squadron RAuxAF arrived, but disbanded along with the other RAuxAF units on 10 March 1957 leaving Church Fenton inadequate for the more modern jet fighters about to enter service. The station's long association with Fighter Command finally came to an end in July 1959, when Flying Training Command took over, and the base continues in the training role today.

Coltishall, Norfolk;
TG270225, 9 miles N of Norwich near B1150

Coltishall was originally built as a bomber base with construction commencing in February 1939, although by this time the RAF Expansion Scheme had been reduced in specification due to the increased costs. Instead of the planned five 'C' Type hangars, only four were built to a reduced height, there was a simpler design of Station HQ facing the main gate and a lower standard of barrack blocks. With the start of

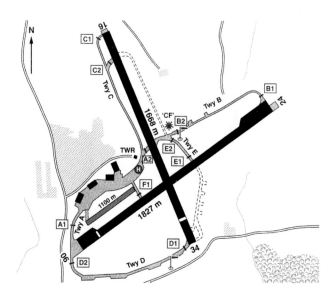

Above: Church Fenton was opened in June 1937 as part of the RAF Expansion Scheme and two runways were constructed during World War 2, which were extended considerably for the operation of Meteors which arrived in October 1950. The Meteors were withdrawn in March 1957 and the station transferred from Fighter to Training Command in July 1959, and the airfield is currently a satellite to RAF Linton-on-Ouse. *Pooley's Flight Guides*

Above: In 1952 Coltishall was developed from its World War 2 configuration including an extended concrete main runway to cope with RAF jet fighter operations and ORPs at both ends. This runway was later extended further to the northeast for Lightnings, resulting in closure of the lane which provided an excellent vantage point for viewing aircraft operations. The main camp was on the western boundary with four utility 'C' Type hangars and large ASPs on either side of the Air Traffic Control. *RAF Museum*

23 Squadron at Coltishall, but because of a high accident rate, particularly on night approaches, and without ejector seat, it was the only unit to operate the early Venom night fighters. The improved NF.3s replaced the earlier night fighters with 23 Squadron and 141 Squadron also received the NF.3s. In 1957 both squadrons were re-equipped with Javelin FAW.4s to form the first Javelin Wing in Fighter Command. In January 1958 141 Squadron was renumbered 41 Squadron and after re-equipping with Javelin FAW.8s departed Coltishall, leaving 23 Squadron which flew Javelin FAW.7s and then FAW.9s until March 1963, when they departed.

Above: The improved Venom NF.3s of 141 Squadron shared Coltishall with 23 Squadron, their first Venom night fighter being delivered in June 1955. At the far end of the line-up is a pair of replacement Javelins. *de Havilland*

World War 2 the security situation changed and in May 1940 the decision was made for Coltishall to become a fighter base with the Spitfires of 66 Squadron arriving on 29 May for the protection of coastal shipping. The station opened formally as a fully operational base within 12 Group Fighter Command on 23 June 1940, by which time the remnants of 242 Squadron had returned from the Battle of France to be commanded by Sqn Ldr Douglas Bader. Coltishall continued to be active during the Battle of Britain, participating in the 12 Group Duxford Wing as well as accommodating squadrons on rest tours from 11 Group, which were allocated to local defence and shipping patrols. The closeness of the station to the North Sea resulted in air-sea rescue (ASR) flights being formed to assist downed airmen, duties which continued many years after the end of World War 2. The airfield was damaged on a number of occasions by the Luftwaffe, and in March 1941 a detachment of 151 Squadron Defiant night fighters was based at Coltishall, a role that was to be a speciality there for nearly 30 years. In June 1941 day offensive operations were started over occupied France, although the airfield was too distant from Normandy to provide any support for the D-Day landings on 6 June 1944.

At the end of the war, the station became home for the Polish Allies with 307 Squadron Mosquito NF.30s and the Mustangs of 303 and 316 Squadrons, until the last of the Poles departed in January 1947. By May 1948 the original runways were in poor condition and a new concrete runway was constructed ready for September 1950 when the Mosquito night fighters of 23 and 141 Squadrons returned. The jet age came to Coltishall in the autumn of 1951 when 23 Squadron re-equipped with Vampire NF.10s and 141 Squadron accepted eight Meteor NF.11s between July and September the same year. The responsibility for introducing the Venom NF.2s to the RAF was

Duxford, Cambs;
TL460460, 7 miles NE of Royston by A505

Construction of the famous Duxford fighter airfield commenced in 1917 and although much of the building had been completed by the end of World War 1, it was not fully occupied until 1919, but it had officially opened on 1 March 1918. The technical site consisted of three double Belfast Warren Truss Girder hangars with a single unit in between two of the doubles. The station HQ was located just inside the main gate which featured a traditional guard room. The officers' mess and personnel accommodation was built on the north side of the A505, separating it from the airfield on the south side. Additional aircraft accommodation was in canvas Bessonneau hangars located to the west end of the site. Duxford was designated as a fighter base, a role which it continued for its operational career. During the Battle of Britain, Duxford became a sector station within 12 Group Fighter Command and it was 19 Squadron based here that was responsible for the introduction of the Spitfire in July 1938. During the Battle of Britain, the Duxford Wing, led by Douglas Bader, provided support for 11 Group in the front line of the South East of England. With the end of the Battle of Britain, Duxford was away from the action and became comparatively quiet, with the Air Fighting Development Unit (AFDU) resident for trials with new types entering service.

In October 1942 the USAAF arrived at Duxford with Airacobras of 345th Fighter Squadron in the 350th Fighter Group while the AFDU continued trials until March 1943 when they moved to Wittering. Duxford was then taken over

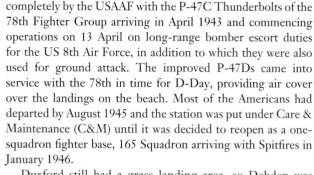

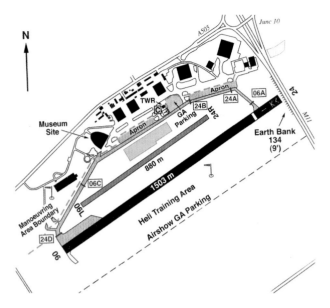

Above: With construction started during WW1, Duxford opened in 1919 as a fighter station, and was a sector station in 12 Group during the Battle of Britain. The main technical site with three double and one single Belfast hangars was located on the northern boundary of the airfield with the accommodation and officers' mess on the other side of the A.505 road. The airfield was used by the USAAF from October 1942 until August 1945, when the last Americans returned home. Fighter Command Meteors arrived in late 1948 and a metal runway was laid by June. However this was insufficient for jet fighter operations and a single 6,000ft concrete runway was constructed from September 1950 until May 1951 when Hunters and Javelins arrived. Hunter and Javelin flying ceased in March 1961 and the station closed in July. Following the making of the film The Battle of Britain in 1968, the site was taken over by the IWM. When the M11 motorway was built at the eastern end a section of the runway was lost, reducing the overall length. *Pooley's Flight Guides*

completely by the USAAF with the P-47C Thunderbolts of the 78th Fighter Group arriving in April 1943 and commencing operations on 13 April on long-range bomber escort duties for the US 8th Air Force, in addition to which they were also used for ground attack. The improved P-47Ds came into service with the 78th in time for D-Day, providing air cover over the landings on the beach. Most of the Americans had departed by August 1945 and the station was put under Care & Maintenance (C&M) until it was decided to reopen as a one-squadron fighter base, 165 Squadron arriving with Spitfires in January 1946.

Duxford still had a grass landing area, so Debden was used as a satellite where its runways were particularly useful, especially when 66 and 92 Squadrons re-equipped with Meteor IIIs from late 1946, and they were joined by 56 Squadron also with Meteors on 17 April. With the entry into service of the more powerful Meteor 4s a metalled runway was laid with working up on the new jet fighters being completed at alternative airfields, before returning to Duxford in June 1948. The airfield was still unsuitable for fast jets, the new plan being to accommodate four squadrons of Mosquito night fighters, but after evaluation, there were insufficient support facilities. Revised plans in 1949 called for the construction of a 2,000-yard concrete runway, the Meteor squadrons departing in October, leaving Duxford on C&M due to funding for the new runway not being available until the following year. The contract for the construction was let in May 1950 with work starting on 18 September 1950 ready for completion in May 1951. In addition to the new runway, enlarged hardstandings were constructed with a T.2 hangar erected at the eastern end of the existing hangars. Aircraft dispersals were allocated at each end of the airfield on both sides of the runway to accommodate each squadron. At each end of the runway were fighter ORPs ready for QRA operations. Both 64 and 65 Squadrons were re-equipped with Meteor F.8 short-range interceptors and arrived at Duxford in August 1951. These aircraft would have had little chance against any Soviet raiders, but all Fighter Command squadrons took turns to stand by fully armed at instant readiness to intercept any potential hostile aircraft. At that time the main threat was from Tu-4s, a Soviet copy of the American B-29, which was believed to be capable of delivering nuclear

Left: No 65 Squadron operated Hunter F.6s at Duxford from November 1956 until disbanding on 31 March 1961. Hunter F.6 M:65 Squadron was being prepared for its next sortie. *J. J. Halley*

Horsham St Faith, Norfolk;
TG220138, N of Norwich, E of A140

Above: No 64 Squadron was stationed at Duxford from 15 August 1951 until moving to Waterbeach on 17 July 1961. The squadron operated Javelin FAW.9s at Duxford from July 1960 and one example was XH708 P:64 Squadron and armed with four Firestreak air-to-air missiles. *J. J. Halley*

weapons. By September 1951 the Duxford squadrons were participating in the quick reaction alert scheme, which continued during the duration of the Cold War, and is still in operation today with the RAF.

The intention was to replace the Meteors of 64 and 65 Squadrons with Hunters, and move them to Horsham St Faith near Norwich, bringing the Fairey Fireflash missile-armed Swifts of 74 and 245 Squadrons into Duxford. The greater-range Swifts would be able to intercept the Russian Tu-4s, but they suffered from serious development problems, and 64 and 65 Squadrons remained at Duxford with the inferior-range Hunters. From August 1956 64 Squadron was switched to the night-fighter role along the lines of other bases fulfilling round-the-clock all-weather capability, their Hunters being replaced by Meteor NF.14s, later to be replaced by the more effective Javelin FAW.7s from September 1958. Meanwhile 65 Squadron received Hunter F.6s in January and February 1957 and the ultimate Javelin FAW.9s with partial reheat began replacing the earlier FAW.7s with 64 Squadron in July 1960.

With the V-Force able to provide a credible nuclear deterrent, Fighter Command was drastically reduced and 65 Squadron was disbanded on 31 March 1961. Duxford suffered from an aging infrastructure and any modernisation was deemed unnecessary, resulting in the station closing to the RAF with the last flight out by a Meteor T.7 on 31 July 1961.

However, this was not the end of flying operations at Duxford, as it was one of the main locations for the 1968 Battle of Britain film where the single Belfast hangar was demolished as part of the effects of a Luftwaffe attack. Following a further period of inactivity, Duxford became the home of the Imperial War Museum housing more than 150 historic aircraft in the surviving Belfast hangars, in addition to the American Air Museum and AirSpace in their own specialist buildings.

Like nearby Coltishall, Horsham St Faith was originally planned as a bomber station, construction starting in 1939 to a similar austere standard as Coltishall with five 'C' Type aircraft hangars without the usual trimmings. The barrack blocks and other buildings had flat roofs rather than the more traditional pitched type, with the main technical site on the SE corner of the station. The station remained quiet until the arrival on 10 May 1940 of 264 Squadron with Defiants and the Station HQ opened on 1 June 1940. During the period of the Battle of Britain, Horsham was the base for Blenheim light bombers of 114 and 139 Squadrons, switching to night intruder operations into 1941. At the end of 1941 105 Squadron arrived with Mosquito B.IVs specialising in low level attack, and were joined by 139 Squadron which re-equipped with Mosquito bombers in June 1942. The Mosquitos left for Marham in September 1942 to allow preparations to be made for the arrival of the USAAF. The 319th Bomb Group passed through Horsham briefly in October 1942 before leaving for North Africa, and the airfield was out of use for flying until April 1943 when 56th Fighter Group arrived with P-47C Thunderbolts, operating on bomber escort missions. The P-47s moved to Halesworth in July 1943 to permit the construction of three runways at Horsham, which was completed with dispersals ready for reopening in January 1944 for the B-24s of the 458th Bomb Group. They completed 240 operations by 25 April 1945, soon after departing back to the USA.

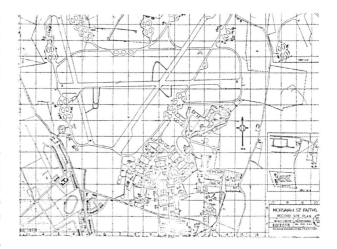

Above: In 1962 Horsham St Faith featured three runways with ORPs at the ends of two of them. The wartime dispersals were still around the perimeter track and the technical site was on the southeast corner with five austerity 'C' Type hangars and an ASP to the north of the hangars. The airfield was mainly used for the second-generation RAF jet fighters, such as Hunters and Javelins before closing to RAF flying on 1 August 1963, and being developed as Norwich Airport. *RAF Museum*

Left: A pair of specially modified Meteor F.8s and a single F.4 of 245 Squadron based at Horsham St Faith undertaking air-to-air flight refuelling trials from a USAF B-29 using the British developed hose and drogue system. *Charles E. Brown*

Fighter Command took over Horsham St Faith on 19 July 1945 with 307 (Polish) Squadron being the first to arrive with Mosquito NF.30s in August, later joined by Mustangs of 64 Squadron and Spitfires of 118 Squadron, making the station very busy and crowded. Following a period when 64 and 65 Squadrons operated Hornets, Meteor IIIs of 74 and 245 Squadrons arrived in August 1946 to form the Horsham Wing, which was completed with the arrival of 263 Squadron in September and augmented by 257 Squadron in April 1947. The squadron began to gradually re-equip with the improved Meteor Mk.IVs from 21 December 1947 and during 1950 these were replaced by the ultimate Meteor F.8 day fighter, going to 245 Squadron from July and 74 Squadron from October. To increase the limited range of the Meteors, some were fitted with flight refuelling probes, giving up to four hours' endurance on convoy patrol duties, the fuel being dispensed from Lincoln tankers. In December 1951 the station establishment was changed to two day-fighter units consisting of 245 and 74 Squadrons still with Meteor 8s, joined by 23 Squadron on 15 January 1952 with Mosquito NF.36s, the ultimate night fighter version of this aircraft to operate with the RAF. The two day-fighter squadrons had their complement of Meteors increased from 16 each to 22, in the hope that any Soviet attack could be swamped by numbers, destroying the entire enemy bomber fleet before a nuclear weapon could be dropped on Britain. On 4 July 1952 23 Squadron left for Coltishall, leaving Horsham as a day-fighter base, the Horsham Wing Meteor F.8s playing a major part in the defence of East Anglia through to 1955.

It was planned that the Supermarine Swift would undergo intensive work-up with 74 Squadron at Horsham, the swept-wing jet having better endurance over the early Hunters. However, the development of the Swift did not go well, and a batch of the early aircraft was issued to 56 Squadron at Waterbeach, who were tasked with trying to solve the problems. In common with many other fighter stations, work commenced at Horsham on a range of improvements in 1954 including a large Aircraft Servicing Platform (ASP) with ORPs at each end of the two main runways, where telescramble was installed. Two protected dispersal areas were established with accommodation for the pilots and maintenance personnel.

In June 1955 245 Squadron moved to a new Wing at Stradishall, and 23 Squadron returned to Horsham in October 1956 with Venom NF.3s. In March 1957 74 Squadron replaced its antiquated Meteors with Hunter 4s, which were in turn replaced by the more effective Hunter 6s in November 1957.

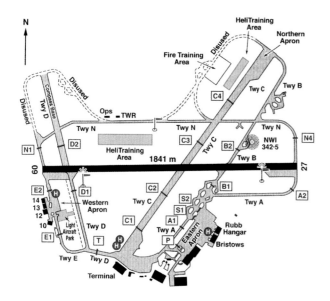

Left: Construction of Horsham St Faith commenced in 1939 as a bomber station with the technical site in the SE corner including five "austerity" 'C' Type hangars. Operations started in May 1940 and the three runways were constructed from July 1943 until January 1944, after which B-24s of the USAAF operated until April 1945. Fighter Command took over in July 1945 and the first Meteor jet fighters arrived in August 1946. In 1954 a large ASP and ORPs at each end of the two main runways were constructed, and the Meteors were replaced by Hunters. The airfield closed to flying in August 1963 and the RAF station closed in March 1967. The site was purchased by Norwich Council and opened as Norwich Airport in May 1970 with a new terminal building at the end of one of the disused main runways. The ASP is now the eastern apron and commercial flights operate from runway 27/09. *Pooley's Flight Guides*

Above: As the night fighter element at Horsham St Faith, 23 Squadron operated Javelin FAW.9s from 31 March 1960 before moving to Coltishall on 11 July 1960. Horsham-based Javelin FAW.9 XH890 M:23 Squadron was armed with four IR heatseeking Firestreak missiles. *J. J. Halley*

As the first Lightning unit, 74 Squadron moved to Coltishall in August 1960, and 23 Squadron replaced its Venoms with Javelins from April 1957, leaving the station the following month. With the reduction in operations, Horsham closed to flying on 1 August 1963, and was finally vacated by the RAF on 24 March 1967. Soon after closure, Norfolk County Council and Norwich City Council decided to purchase the airfield site for development as a regional airport, and on 30 May 1970 it reopened for flying as Norwich Airport, which still continues.

Leconfield, East Riding of Yorkshire;
TA030435, N of Beverley on side of A164

Leconfield was constructed as a bomber base in the RAF Expansion Scheme in the mid-1930s. Located about two miles north of Beverley, the site was restricted by the Beverley to Driffield railway line on the eastern boundary and the A164 on the western side. The technical site was built alongside the A164 with five 'C' Type hangars in a semicircle, and behind were workshops, stores, transport, administration, messes and living accommodation. The station opened on 3 December 1936 as part of 3 Group, Bomber Command, and the first flying residents were the Heyfords of 166 Squadron, which arrived on 7 January 1937 from Boscombe Down, joined a few weeks later by 97 Squadron. After these units moved out in September 1939, the airfield was reduced to C&M until it became part of 13 Group Fighter Command and Spitfires of 616 Squadron arrived in October 1939. For the first two years of World War 2, Leconfield was temporary home to many of the fighter squadrons both in the Battle of Britain and the offensive operations that followed. These squadrons were responsible for local and coastal defence while they replaced lost pilots and retrained new personnel.

In 1942 the station was developed as a heavy bomber base with three concrete runways linked by a perimeter track with bomber dispersals. The first arrivals were Wellington-equipped 196 and 466 Squadrons from Driffield, the first operation being by 466 Squadron on the night of 15/16 January 1943. In September Halifaxes began to replace the Wellingtons with 466 Squadron, which they operated for the rest of the War, and similarly-equipped 640 Squadron was formed on 7 January 1944. On 30 December 1944, 96 Squadron re-formed as part of RAF Transport Command with 25 Halifax VII transport versions of the heavy bomber. The aircraft were configured as troop transports for flights to India and returning casualties to Britain, Cairo being the main staging post. After only a few weeks at Leconfield, the plans were changed and the squadron was sent to its new base at Cairo. The final bombing mission of World War 2 from Leconfield was on 25 April 1945 when both 640 and 51 Squadrons bombed gun emplacements.

Above: In 1952 Leconfield was developed considerably from its World War 2 configuration for the operation of RAF jet fighters. All three runways were extended, with a later extension to the south of the main runway. Dispersals around the perimeter track were retained with the main camp on the western side of the landing ground including five 'C' Type hangars and a dispersed site on the southeast boundary. Following the final departure of 19 and 92 Squadron Lightnings to Germany at the end of 1965, 60 MU were residents for a while and the camp was taken over by the Army on 1 January 1977. *RAF Museum*

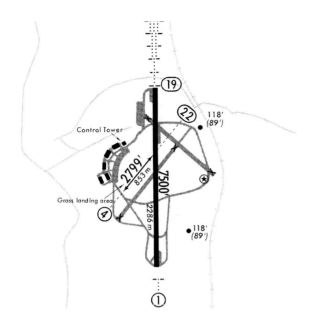

Above: The May 1969 airfield chart of Leconfield shows its ultimate configuration after the Lightnings had departed, with the main runway extended to 7,500 feet with ORPs at both ends and the other two runways disused. The technical site with five 'C' Type hangars was located on the western boundary and after use by 60 MU, the station was closed and taken over by the Army. *Jeppesen*

Leconfield was retained as a permanent station after World War 2 and from October 1957 improvements were carried out on the station including lengthening the main runway in preparation for becoming a frontline fighter base. The first units to arrive were 19 Squadron with Hunters and 72 Squadron with Javelin FAW.4s in June 1959 from Church Fenton. These early Javelins were a challenge for the ground crews to keep serviceable, and although it was hoped that they might be replaced by the ultimate FAW.9, 72 Squadron

Below: No 19 Squadron re-equipped with Lightning F.2s in December 1962 and moved to RAF Germany on 23 September 1965. Lightning F.2 XN730 served as 'B' with 19 Squadron. *Author's collection*

was disbanded on 30 June 1961. Meanwhile the Hunters of 92 Squadron had arrived at the end of May, and both 19 and 92 Squadrons converted to be the only operators of the Lightning F.2s. Leconfield was not to remain a frontline Cold War fighter station for much longer, with 19 Squadron moving to Gutersloh in West Germany on 23 September 1965, followed by 92 Squadron to Gelsenkirchen on 29 December. The station then became 60 MU for a few months until final closure and was taken over by the Army on 1 January 1977 as Normandy Barracks.

Leeming, North Yorkshire;
SE305890, W of Northallerton, alongside the A(1)M

The site, having been used as a flying club in early 1938, was requisitioned by the Government in late 1938 for the construction of a standard RAF bomber airfield with the technical site located in the south-east corner. The five main hangars were the austerity version of the 'C' Type to save costs and the station opened in June 1940 with 10 Squadron arriving the following month with Whitley Mk.V bombers, and the first operation was on 20 July. With the continued build-up of the bomber force, 7 Squadron re-formed on 1 August 1940 with Short Stirlings, the first unit in the RAF to operate four-engine heavy bombers and after work-up moved south to Oakington in October. In November 35 Squadron, which had re-formed at Boscombe Down on 5 of the month, had the task of introducing the new Handley Page Halifax to RAF service, moving to Linton-on-Ouse in early December. Meanwhile 10 Squadron continued to operate the veteran Whitleys, converting to Halifaxes by the end of 1941. The resident bomber units continued the campaign against the enemy with the Tirpitz attracting some attention, but without success. On 1 January 1943, Leeming was transferred to 6 Group RCAF where the Dominion squadrons formed up and were trained in the operation of heavy bombers. As things settled down, 427 (Lion) Squadron RCAF arrived on 5 May 1943, where they were based for the next three years, and in August they were joined by 429 (Bison) and 408 (Goose) Squadrons, all three carrying the offensive over Germany. To cope with operations by heavy bombers, three runways had been laid, connected by a perimeter track with bomber dispersals for the aircraft when they were not on heavy maintenance in the hangars. The last offensive operation by both Lion and Bison squadrons, by then flying Lancasters, was on 25 April 1945, following which they remained in Britain to help return Allied prisoners of war and British troops back from Italy before disbanding on 1 June 1946.

During 1956 repairs were made to the runways including a major extension to the southern end to accommodate Gloster Javelins which arrived for 228 OCU in June 1957. In September 1957 264 Squadron arrived from Middleton St George, but was disbanded on 1 October, 33 Squadron reforming the same day

Above: No 60 Squadron arrived from Tengah in Singapore on 27 May 1959 to convert to Meteor NF.14s from October 1959, which it operated until August 1961. The aircraft are on the Leeming flight line with WS794 K:60 Squadron nearest. *RAF Museum PC73/4/549*

Leuchars, Fife;
NO460205, 5 miles NW of St Andrews by A919

with Meteor NF.14s which they operated until replaced by Javelins in July 1958, when the squadron moved to Middleton St George. With the task of training all the Javelin crews, 228 OCU began the first course in October 1957 until August 1961 when the OCU was prematurely disbanded in the mistaken belief that enough crews had been trained to sustain the Javelin force. Leeming then took on a training role, so its contribution to the early part of the Cold War was minimal. However, it did participate in the later strategic part of the Cold War, as the base for 25 Squadron Tornado F.3s.

Aviation first came to Leuchars in 1911 in the form of balloon experiments by the Royal Engineers, a formal airfield being established in 1918 by the Royal Naval Air Service (RNAS), alongside the estuary of the River Eden. The original camp was very compact with three double Belfast hangars alongside the landing ground, with a single Belfast hangar situated behind, surrounded by the workshops, stores, administration and accommodation buildings. The station continued in the naval aviation role forming squadrons and flights for service aboard RN ships, until 1938 when the station was taken over by Coastal Command, with 224 and 233 Squadrons equipped with Ansons and later Hudsons which arrived in 1939. During the early part of World War 2, the station was expanded considerably with four new utility 'C' Type hangars to the west of the original technical site and new administration and technical buildings to the rear. Two of the original double Belfast hangars were retained, but the middle one and the old buildings to the rear were gradually replaced with new structures. Two concrete runways were also constructed to allow all-weather operations with the larger aircraft. While continuing in the coastal role, 540 Squadron was formed in October 1942 from the Reconnaissance Unit (PRU) equipped with unarmed Spitfires and Mosquitos, their primary task being to monitor German Navy activities. Leuchars became an anti-shipping strike base from January 1943, as well as the home of the BOAC-operated "Ball Bearing" flights from Stockholm from 1941, initially with Hudsons, and then Mosquitos from

In addition to being a fighter station, Leeming was also used by Training Command. Vampire T.11s were used by 3 FTS to train foreign students and were finally withdrawn from RAF flying training on 29 November 1967. The final line-up is in front of the typical post-World War 2 air traffic control tower with austerity 'C' Type hangars in the background. *Author*

February 1943. The BOAC Stockholm services were moved to Croydon in May 1945 after a total of 520 round trips. The final role of Leuchars in World War 2 was still coastal work with Liberators of 206 Squadron on anti-submarine duties, their last operational sortie being on 3 June 1945.

Following a period as a training station, the change was made to the fighter role when the jet age arrived in May 1950 in the form of 222 Squadron equipped with Meteor 4s from Waterbeach. They were joined by 43 Squadron from Tangmere in November. By the end of 1951 both 43 and 222 Squadrons had been re-equipped with Meteor F.8s and they were joined by 151 Squadron on 15 September with Vampire NF.10s, making up the Leuchars Wing.

The resident day fighter units began to re-equip with Hunter F.1s on 29 July 1954 starting with 43 Squadron, followed by 222 Squadron, both at Leuchars. It was during the introductory service development flying with the Hunter F.1s that restrictions were made in the firing of the 20mm cannons at high altitude, causing the Avon engines to flame-out. This was caused by tight turns during high altitude interceptions, producing shock waves entering the air intakes, in addition to the exhaust gases from the gun firing. The Hunter F.1 was therefore unsuitable for its intended role as a high-altitude interceptor, although the Sapphire powered Hunter F.2 did not suffer from the same flame-out problems. Both the Hunter F.1s and F.2s suffered from a very limited endurance and as a result only five Fighter Command squadrons were equipped with the aircraft. Despite its limitations it was useful as an interim type taking over from the antiquated Meteors, until the arrival of the more effective Hunter F.4s and F.5s By this time 151 Squadron was flying Meteor NF.11s, and converted to Venom NF.3s.

Above: Leuchars-based No 23 Squadron exchanged its Javelin FAW.9s with Lightning F.3s between August and October 1964, and each type was represented in this formation. *RAF Museum P015110*

During the summer of 1956, the airfield was closed for the main runway to be extended to the east, the home-based squadrons returning in August. No 151 Squadron re-equipped with Javelins in 1957, but 222 Squadron disbanded on 1 November, and more Javelins arrived with 29 Squadron from Acklington joining 151 in the all weather defence of the North of Britain. The Hunters of 43 Squadron left for Cyprus in June 1961 and 151 was disbanded on 19 September to be replaced by 25 Squadron with Javelins in December, which in turn disbanded on 30 November 1962. In its place the Javelins of 23 Squadron arrived in February 1963, and in the following month 29 Squadron departed for Cyprus. The replacement of the second-generation jet fighters commenced at Leuchars with the arrival of 74 Squadron with Lightnings on 2 March 1964 which moved from Coltishall, and 23 Squadron exchanged its Javelins for Lightning F.3s in August 1964. Javelin operations continued at Leuchars for a period with the re-formed 228 OCU on 1 May 1965 to train new crews for the Near and Far East theatres. Leuchars then became an essential strategic fighter defence base for the northern approaches to Britain by potential Soviet raiders.

Left: No 151 Squadron operated Venom NF.3s from Leuchars from September 1955 until June 1957 when they converted to Javelins. Venom NF.3 WX794 had an accident at Turnhouse in May 1957, following which it was relegated to the dump. *Author's collection*

Linton-on-Ouse, Yorkshire;
*SE490617 NW of York,
close to the north bank of the River Ouse.*

This station was constructed as one of the earliest bomber bases in the RAF Expansion Scheme during the mid 1930s. The technical site was located in the south east corner of the airfield with a semicircle of four 'C' Type hangars, and a fifth behind the southerly one. Behind these hangars was a substantial group of workshops, stores, administration buildings, barracks and messes in the standard style of the time. So similar were the bases, that anyone moving from one to another would have little difficulty finding their way around. The landing area featured three runways with bomber dispersals around the interlinking perimeter track and bomb dumps located in the shelter of nearby woodlands. The station was officially opened on 13 May 1937 while construction was still under way, the initial occupants being the staff of 4 Group HQ. The first aircraft to arrive were Whitleys of 51 and 58 Squadrons and immediately on the outbreak of war these two units were the first to fly over Germany during World War 2, their task being to drop leaflets. With the build-up of operations and squadrons at the base, 4 Group HQ moved out. Halifax heavy bombers supplemented the vulnerable Whitleys when 35 Squadron arrived from Leeming on 5 December 1940 and after working up with the new type commenced operations on the night of 10/11 March 1941 with the bomber offensive continuing, both squadrons attacking targets in Germany and occupied France. The Whitleys of 58 Squadron were finally withdrawn following an attack on Ostend docks on the night of 29/30 April 1942. When the Pathfinder Force was formed in August 1942, 35 Squadron was one of the first five squadrons to join, and left for Graveley. On 16 September 76 and 78 Squadrons arrived at Linton flying Halifax B.IIs, 76 Squadron being commanded by Group Captain Leonard Cheshire. Their stay was short as in October 1942 the HQ of No.6 Group RCAF was formed at Linton, and in mid-June 1943 both resident squadrons moved out to be replaced initially by 426 (Thunderbirds) Squadron RCAF and began working up on Lancaster B.IIs. They were ready for the first operation on the night of 17/18 August as part of the force attempting to destroy the experimental Peenemunde facility. On 28 August 1943, 408 (Goose) Squadron RCAF arrived from Leeming initially with Halifaxes, but in October converted to Lancaster IIs. The two squadrons contributed to the Battle of Berlin from mid-November 1943, and in early 1944 the Lancaster IIs were phased out and replaced by Halifax B.IIIs, and then Halifax B.VIIs until the end of the war, when both squadrons made their final attacks on 25 April 1945, the last Canadians departing on 5 September.

Linton was returned to the RAF in June 1945 with a transfer to Fighter Command in July 1946, the first unit to arrive briefly being 264 Squadron with Mosquito NF.36s. During the first half of August 1946 the Hornet-equipped 64 and 65 Squadrons arrived from Church Fenton which were replaced by Meteor F.8s in October 1949. The two resident squadrons were joined by Meteor-equipped 66 and 92 Squadrons from Duxford. In August 1951 64 and 65 Squadrons moved out to Duxford and at the end of the month 264 Squadron moved in from Coltishall, replacing its Mosquito NF.36s with Meteor NF.11s in November 1951, and the more advanced NF.14s in October 1954. The squadron moved to Middleton St George in February 1957.

Linton-on-Ouse was unique in RAF Fighter Command as being the sole base for Canadair Sabres, the remainder of the type serving with the 2nd TAF in Germany. The Sabre was selected as a stopgap swept-wing fighter to overcome the failure of the Swift and delays in the development of the early Hunters. In early 1952 an order was placed for 370 Sabres, of which all except three would be Canadian-built Sabre 4s, the remaining three being USAF F-86Es.

Left: The F-86 Sabre Mk.4 was a Canadian-produced version of the North American Sabre. The type was used by the RAF at Linton-on-Ouse as an interim swept-wing jet fighter pending the delayed deliveries of the Hunters. Two squadrons operated the type in Britain and further squadrons were operational in Europe. The first Sabres arrived at Linton-on-Ouse for 66 and 92 Squadrons, and the type was finally withdrawn in June 1956.
Crown Copyright

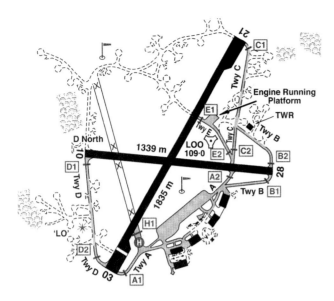

Left: Linton-on-Ouse was built in the mid 1930s as an RAF Expansion Scheme bomber station with five 'C' Type hangars in the technical site located in the southeast corner. The station opened for flying in May 1937 and three runways were constructed for heavy bomber operations by the RCAF from June 1943. The Canadians left by September 1945 and Fighter Command took over in July 1946, with Meteors arriving in October 1949. Two of the runways were extended for RAF Sabre 4 operations which was from January 1954 to May 1956, when they were replaced by Hunters. On departure of the Hunters the station was taken over by RAF Training Command, which continues with Tucanos providing basic pilot training. *Pooley's Flight Guide*

An additional 60 were also initially loaned to the RCAF in Europe until 1 December 1953, when they were passed to the RAF, making a total Sabre fleet of 430 aircraft. The first ten Sabres were scheduled for acceptance by the RAF in December 1952, with completion of delivery in October 1953, and all the aircraft were to be flown across the North Atlantic by 147 Squadron ferry pilots, the entire operation being named Bechers Brook after the famous horse jump on Aintree racecourse. The first three Sabre F.2s to be used for the training of pilots and ground crews left Canada on 28 September 1952 and arrived at North Luffenham on 10 October, a total of 60 jet-experienced pilots being selected. The first ferry mission consisting of nine Sabres left Canada on 9 August 1953, flying via Goose Bay, Bluie West 1 in Greenland, Keflavik in Iceland and Prestwick in Scotland. Unfortunately one Sabre was lost on the approach to Prestwick, killing the pilot. The normal number of Sabres flying across the North Atlantic was a batch of 30 aircraft, and the final deliveries were completed in May 1954. The initial deliveries were to 2nd TAF Squadrons in Germany, No. 66 Squadron converting from Meteor F.8s at Linton-on-Ouse in December 1953, followed by 92 Squadron in February 1954.

The first Sabres for 66 Squadron were delivered from 5 MU at Kemble on 27 November 1953, but due to protracted acceptance checks, it was some time before they were available for the squadron pilots. Further Sabres arrived from 14 January 1954 to allow the conversion courses to start, the initial experienced pilots coming from 147 Squadron. There were enough pilots qualified to participate in Exercise Magna Flux during early March when they were tasked as defending fighters against simulated raids from NATO aircraft based in Europe. The last Sabre for 66 Squadron was delivered on 8 March making a total of 24 aircraft, but as the correct aircraft strength should have been 22 aircraft, two were transferred to 92 Squadron to begin their conversion from Meteors.

All the pilots of 66 Squadron had been converted from Meteors to Sabres by 1 April, the main role being air-to-air gunnery. In July 1954, 66 Squadron was tasked with air defence of the UK as part of Exercise Dividend, testing the defences against high- and very high-altitude attack which would be likely in the early stages of a nuclear war. The 'attacking' force included USAF B-47 Stratojets, RCAF F-86s and Thunderjets, but the high-level B-47 formations proved to be virtually impossible to intercept, as their speed was similar to the fighters of the day. Including the Linton Sabres, the UK fielded some 877 fighters during the exercise, as well as USAF F-86Ds plus Vampires, Venoms, Meteors and RCAF Sabres from North Luffenham. Around 30 percent of the attacking aircraft were intercepted before crossing the British coastline, but there was a shortage of all-weather fighters in the Exercise. The Sabres were more successful at practice intercepts of Canberras, and during sector exercises in March 1955 the Sabres were given the opportunity of intercepting the massive B-36 bombers at altitudes of up to 43,000 feet, which were found easier to intercept than the Canberras. The Sabres were not always able to catch up with the slow-flying giant, often due to lack of endurance.

The Sabres operated in the annual air defence Exercise Beware in late September 1955, with 66 Squadron flying 174 sorties against Vampires, Meteors and Canberras at heights between 50 feet and 51,000 feet, claiming 80 successes, but on detachments to Lossiemouth and Kinloss in November, the Sabres were successfully intercepted by Hunters of 43 and 222 Squadrons from Leuchars. This clearly demonstrated the decline of the Sabre and towards the end of the year a number of the squadron personnel began to receive instruction on the Hunter. The replacement of the Sabres by Hunters commenced in March 1956, although during the build-up, to keep pilots current, some Meteor F.8s were also supplied, allowing a mixed formation on 19 April with a Hunter leading a Sabre and Meteor in echelon to starboard. The last Sabres departed on 16 May.

Meanwhile the Meteor F.8s of 92 Squadron were replaced by Sabre F.4s in February 1954 and the pilots were introduced to new items of flying clothing including anti-g suits and bone-dome flying helmets. The first Sabre arrived on 23 February from 33 MU at Lyneham, and the full squadron strength of

Above: Linton-on-Ouse was the main RAF station for the Sabre F.4s in the UK. The Sabre F.4s were with 66 Squadron from January 1954 until March 1956, when they were replaced by Hunter F.4s. Due to a shortage of Hunters initially, Meteor F.8s were returned temporarily which allowed this mixed formation in March 1956. *RAF Museum P015107*

Middleton St George, County Durham;
NZ375130 6 miles E of Darlington by B1273.

This station was one of the mid-1930s RAF Expansion Scheme bomber bases, but work on construction did not start until 1939, just before the beginning of World War 2. As a result, materials were in short supply and only two permanent hangars were built on the technical site located on the NW side of the airfield. These hangars consisted of one 'C1' and a 'J' Type, which were later supplemented by a pair of T.2s and a 'B1'. Middleton St George was officially opened on 15 January 1941 as part of 4 Group, Bomber Command and during April 78 Squadron arrived from Dishforth with Whitley B.Vs. They were joined by 76 Squadron with Halifax B.Is, only the second unit to operate this new heavy bomber, and both squadrons moved to Linton-on-Ouse in September 1942. The station was then prepared for the arrival of bomber squadrons of the RCAF with 420 Squadron arriving on 16 September operating Wellington B.IIIs, and 419 Squadron arriving on 10 November. In May 1943 420 Squadron was posted to North Africa and replaced by 428 (RCAF) Squadron on 3 June, the two resident units remaining until after the end of World War 2, flying Halifax bombers, later replaced by Lancasters. The final wartime mission for both squadrons was on 25 April 1945, the operations at Middleton St George being very similar to those at Linton-on-Ouse.

22 aircraft was achieved on 30 March, while training of the pilots was underway. The Sabre was popular with both pilots and ground crew with its advanced technology and spacious cockpit. The flying controls were fully powered, apart from the flaps which were electrically operated. 92 Squadron participated in the same exercises as 66 Squadron, and a Canberra was intercepted successfully at 48,000 feet, close to the maximum ceiling of the Sabre. For Exercise Beware in late September 1955, 92 Squadron deployed to Driffield for a ten-day detachment with 18 Sabres complete with ground equipment. The squadron flew 138 sorties against high-flying RAF aircraft making dummy raids against Britain, the intercepts being made at around 48,000 feet, the limit of the Sabre. A total of 78 claims were made, and serviceability had achieved a creditable 90 percent. From February 1956, ground school for the Hunter began to become established with Sabre flying reducing, but as insufficient Hunters were available, 92 Squadron reverted to Meteor F.8s, while the Sabres were withdrawn prematurely. The Sabres began to depart in April 1956, the last departing on 11 June, with Hunter F.4s replacing the venerable Meteors as they became available. 92 Squadron moved to Middleton St George on 1 March 1957.

With the airfield unsuitable for the operation of high-performance jet fighters then entering service, 66 Squadron had already gone to Acklington in February, and the station was transferred to Training Command. It is still very busy today with fast jet pilot training as 1 FTS on Tucanos, before the students move to Valley for instruction on the Hawks.

Above: In 1963 Middleton St George was considerably enlarged to allow the operation of the future RAF fighter aircraft including Lightnings. The major layout change was to extend the main runway to the southwest with ORPs at each end. The technical and accommodation site was on the northwest side of the airfield with an assortment of B.1, C.1, J and two T.2 hangars, and the station was closed to RAF flying on 17 April 1964. It reopened as Tees-Side Airport on 1 April 1966. *RAF Museum*

Above: The commanding officer of 92 Squadron was Sqdn Ldr R. H. B. Dixon, who is standing by his Hunter F.6 with a Middleton St George T.2 hangar in the background. *RAF Museum P.008875*

With the departure of the Canadians, the station was transferred to Fighter Command, 13 Operational Training Unit arriving with Mosquitos in July 1945, moving to Leeming in May 1947. Although the station was then transferred to Training Command, 608 (North Riding) Squadron RAuxAF arrived in 1947 flying Mosquito NF.30s which were replaced by Spitfire F.22s in 1948. During 1950 the Spitfires were replaced by Vampire F.3 jet fighters, which in turn were replaced by Vampire FB.5s in 1952. As with all the other RAuxAF units, 608 Squadron was disbanded in March 1957. Middleton St George returned to Fighter Command in June 1956 and 92 Squadron flying Hunter F.6s arrived in March 1957 and were joined by 264 Squadron Meteor NF.14s. Both squadrons remained until September, when 92 Squadron moved to Thornaby and 264 moved to Leeming where it was renumbered 33 Squadron.

The airfield then closed for major improvements including lengthening of the main runway to 7,500 feet to accommodate the new generation of supersonic fighters. By then equipped with Javelin FAW.7s, 33 Squadron arrived in July 1958 and was joined by 92 Squadron, who were Fighter Command's official aerobatic team, named "The Blue Diamonds". In May 1961 92 Squadron moved to Leconfield and 33 Squadron, who had converted to Javelin FAW.9s, remained at Middleton St George until disbandment on 17 November 1962. From 1 June 1963 the station was home to the Lightnings of 226 OCU, but these moved to Coltishall on 13 April 1964 when the station closed on 17 April, even though it had been an RAF Master Diversion airfield, open round the clock for any RAF emergencies. The airfield was then opened as Tees-Side Airport on 1 April 1966.

Above: Venom NF.2 WL872 of 219 Squadron at Middleton St George – now Durham Tees Valley Airport – for the Battle of Britain display in September 1956. The squadron was based at Driffield from 5 September 1955 until it disbanded on 31 July 1957. *Eric Taylor*

North Luffenham, Leicestershire;
SK940050, 6 miles W of Stamford by A6121.

This station was built as a rather austere bomber base, construction starting in early 1940, and it served briefly as a Cold War airfield with the F-86E Sabres of the RCAF. The technical site consisted of two 'J' Type hangars with the technical and domestic buildings having flat roofs, and at the time of opening on 18 January 1941 the grass airfield was in use by Tiger Moths of 17 EFTS. The first bombers arrived in the form of Hampdens of 61 and 144 Squadrons in July 1941

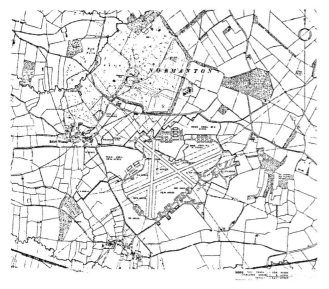

Above: North Luffenham was built as a very austere station with a three-runway layout and technical sites on the north and west sides of the airfield. The westerly site consisted of two 'J' Type hangars with the site on the northern boundary having a B.1 and three T.2 hangars. The main active flying use of the station was by the F-86 Sabres of the RCAF 1 Fighter Wing from 1 November 1951 with 410, 439 and 441 Squadrons which stayed until moving to Germany by 1 April 1955. For these operations ORPs were added at both ends of the main runway and ASPs by each group of hangars. The station is now occupied by the 16th Regiment Royal Artillery. *RAF Museum*

Above: When flying ceased at North Luffenham in March 1958, the station was allocated as a Thor missile base, the first arriving in October 1959. The missiles were withdrawn in September 1963, and the station continued with MOD non-flying activities, until taken over by the Army. *IWM RAF T.2611*

which continued on bomber operations until moving out in April 1942. The station was then transferred to the training role and from mid-1943 major improvements were made to the base including the erection of a 'B1' and three T.2 hangars on the northern edge of the airfield. The reconstruction included a standard three-runway Class A airfield and it reopened in March 1944 in the training role.

On 1 November 1951 No.1 Fighter Wing RCAF arrived and more changes were made to the station, both shorter runways becoming disused while the main one had ORPs on the north side at both ends. Both hangar sites had Aircraft Servicing Platforms (ASPs) constructed to accommodate some 75 Sabres of 410, 439 and 441 Squadrons RCAF, as part of 12 Group Fighter Command. 410 (Cougar) Squadron RCAF was the first to exchange its Vampires for Sabres, followed by 441 (Silver Fox) and the aircraft of both squadrons were cocooned against salt corrosion and shipped across the Atlantic, arriving at North Luffenham by the end of 1951. On 1 September 439 (Tiger) Squadron formed with Sabres and after work-up, to save time, the squadron pioneered the North Atlantic route by air, beginning Operation Leapfrog from Bagotville via Goose Bay, Labrador, Bluie West One in Greenland,

Left: No 1 Fighter Wing RCAF arrived at North Luffenham in November 1951 and consisted of 410, 439 and 441 Squadrons. RCAF F-86E Sabre 19379 carrying the code 'BH' and badge of 430 Squadron RCAF was a visitor to North Luffenham, and the RCAF Sabres departed for Germany by April 1955, when the station was returned to the RAF. *Michael Bowyer*

Keflavik in Iceland and RAF Kinloss before completing the last leg to North Luffenham. Despite delays caused by weather en route, the 21 Sabres arrived safely on 15 June, bringing No.1 Wing to full strength. The major tasks of the squadrons of No.1 Wing, together with the other RCAF Wings which formed the No.1 Air Division was to patrol the East German borders and maintain armed Zulu alert, on standby fully armed to scramble and intercept any unidentified hostile raiders. The Canadians began moving out to European NATO bases at the end of 1954 and on 1 April 1955 the station returned to the RAF, again in the training role, flying ceasing at the end of March 1958. It then became a Thor missile base, reopening on 1 October 1959 until the last departure of the missiles on 27 September 1963. What was a very basic base is now occupied by the 16th Regiment Royal Artillery, which is equipped with the Rapier low-level air defence system.

North Weald, Essex;
TL488044, 2 miles NNE of Epping.

Military flying commenced at North Weald in August 1916 with a detachment of 39 Squadron equipped with Be.2cs used for the air defence of London. The first success was on 1 October 1916 when 2/Lt W. J. Tempest took off to engage night raiders and after nearly two hours in the air he spotted Zeppelin L.31 caught in searchlight beams. He made his attack and shot the airship down over Potters Bar, for which achievement he was awarded a DSO. With the growing Gotha bomber threat, 39 Squadron moved in completely and it was joined by 75 Squadron with Bristol Fighters in May 1918. Following the end of World War 1 flying ceased but

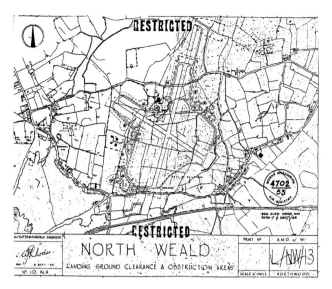

Above: For jet fighter operations North Weald was considerably enlarged in 1953 beyond its World War 2 boundaries with two runways laid early in World War 2, the main one being extended twice to the north while the cross runway was extended to the west. The main technical site was in the southeast corner with two pre-World War 2 'A' Type hangars, one of which has since been destroyed by fire. North Weald was reduced to C&M on 15 November 1958, and adopted for use by general aviation in August 1979. Additional dispersed sites were built with T.2 hangars on the southwest corner with large aprons to house additional squadrons.
RAF Museum

Below: No 604 Squadron RAuxAF operated Vampire F.3s at North Weald from November 1949 until replaced by Meteor F.8s in August 1952. The squadron took part in Exercise 'Emperor' in October 1950. *de Havilland*

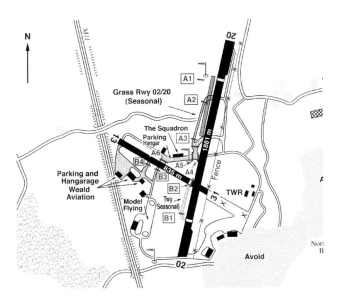

Above: The origins of North Weald go back to World War 1 when it was opened in August 1916 for the air defence of Britain against enemy raiders. The airfield closed in 1919, but reopened in September 1927 with two 'A' Type hangars on the technical site in the southeast corner of the flying field. It became a sector station in 11 Group during the Battle of Britain and two runways were constructed in 1939/1940. In November 1949 Vampire jet fighters arrived and the runways were extended to the north and west to allow operations by Meteors, followed by the Hunters of 111 Squadron. At this time additional T.2 hangars were erected on the western side of the airfield to accommodate the jet fighter squadrons, and the main runway was further extended to the north. The RAF closed the station in November 1958, and following its use by the *Battle of Britain* film production in 1968, North Weald reopened for flying as a general aviation airfield in August 1979. *Pooley's Flight Guides*

reconstruction started in 1926 of a fighter station ready for re-opening to take place on 27 September 1927. The technical site in the south-east corner featured two 'A' Type hangars with sufficient space in between for a third to be built. Included in the stores, workshops and administration accommodation were brick and slate barrack blocks. The first units to arrive were 56 Squadron with Siskins on 11 October 1927, to be joined by similarly equipped 29 Squadron on 1 April 1928, making up a typical two-squadron, twenty-aircraft-per-station strength. Both squadrons re-equipped with Bulldogs in 1932, and the station continued in the fighter role leading to a major participation in the Battle of Britain. The first Hurricanes arrived for 56 Squadron in May 1938, with 151 Squadron receiving their Hurricanes in December. Two paved runways were constructed, some of the earliest at Fighter Command airfields, the north-south one being 2,800 feet long, and the east-west runway 2,750 feet. North Weald was designated as a Sector Station within 11 Group, Fighter Command with Stapleford Tawney as its satellite, and Martlesham Heath as a forward airfield with the station squadrons being very active

throughout the Battle. There was some damage sustained from Luftwaffe bombers, the first major attack coming on 24 August 1940 with severe damage to the officers' and airmen's married quarters, and nine army troops killed in a direct hit on a shelter. This was followed by the heaviest raid on North Weald on 3 September when 200 bombs fell on the station, gutting the hangars and damaging the operations block and many other buildings. Casualties were four killed and 37 injured. The final raid during the Battle was on 29 October when more than forty 50kg bombs were dropped by Bf.109 fighter-bombers, causing considerable damage. The station then went on to the offensive including 71 (Eagle) Squadron of American volunteers and two Norwegian squadrons. With North Weald being distant from the action around D-Day, the level of flying reduced.

Fighter units returned to North Weald on 27 March 1949 with the arrival of Spitfires of 601 and 604 Squadrons RAuxAF. In November 1949 these squadrons began to re-equip with Vampire F.3s, and were joined in March 1950 by 72 Squadron with Vampire FB.5s. To allow the safe operation of these early jet fighters, the main north-south runway was extended, and replacement of the Vampires began in August 1952 with 72 Squadron receiving Meteor F.8s. 604 Squadron followed with Meteors the same month and 601 in September. On 11 May 1953, 72 Squadron departed from North Weald and in October the station became a Sector Operations Centre and HQ of the Metropolitan Sector. To accommodate these new jet fighter squadrons additional T.2 Hangars and hardstandings were constructed on the southwest corner of the airfield and in December 1953, 111 Squadron re-formed at North Weald, initially equipped with Meteor F.8s, to complete the Wing, and in June 1955 it re-equipped with Hunter F.4s. These were replaced by Hunter F.6s from November 1956, and led by Sqdn Ldr Roger Topp 'Treble One' Squadron became the RAF Aerobatic Team, known as the Black Arrows, flying appropriately-painted aircraft. On 30 April 1957 ACM Sir Harry Broadhurst presented the Squadron Standard to 111 Squadron, following which the team was presented to the media with 11 Hunter F.6s on parade, three in standard camouflage and the remainder in overall gloss black. The culmination of formation aerobatics was achieved by 22 Hunters of 111 and 43 Squadrons at the 1958 SBAC Air Show, which included a formation loop in their programme.

With the drastic cuts to Fighter Command, including the disbandment of the RAuxAF squadrons on 10 March 1957, activities at North Weald rapidly declined and on 18 June 1958, Treble One departed and the station was placed under C&M on 15 November 1958. Spitfires returned in 1968 for the making of the Battle of Britain film and in August 1979 the site was bought by Epping Forest District Council as an industrial park and airfield for general aviation. One of the 'A' Type hangars was destroyed by fire, but the other survives in use for storage.

Left: The Vampire F.1s first entered service with 247 Squadron at Odiham, although they also operated briefly at West Malling, Chilbolton and Acklington. The squadron began to re-equip with Vampire F.3s in July 1948. *de Havilland*

Odiham, Hampshire;
SU740491, 6.5 miles ESE of Basingstoke by A32.

Military aviation first came to the Odiham airfield site in 1926 when the Air Ministry purchased Down Farm, a mile south of Odiham town, for use as Summer Camp landing ground. Bessoneaux canvas hangars were erected to shelter the Bristol Fighters of 4 and 13 Squadrons and the rest of the accommodation was under canvas. With the RAF Expansion Scheme of the 1930s, Odiham was selected for development and additional land was purchased with site clearance commencing in 1934 to accommodate an Army Co-operation three squadron station. The technical site was situated in the north-west corner of the landing ground alongside the A32 Alton road with three 'C' Type hangars for aircraft accommodation, with stores, SHQ, administration buildings messes and H-type barracks, all built to the traditional style. The hangars were in a semi-circle, with the centre one set back from the other two. The construction was completed by Christmas 1936 and RAF Odiham was handed over to 22 Group on 3 December as the HQ of 50 (AC) Wing. Both 4 and 13 Squadrons were in residence by mid-February 1937. On 8 April 1938 53 Squadron arrived to complete the Wing, and began to re-equip with Blenheims from January 1939, while the other two units received Lysanders. With constant problems with waterlogging, Odiham became one of the first airfields with concrete runways, three being constructed by the spring of 1939, later to be improved and with interconnecting perimeter track featuring dispersals along the southern side. At the outbreak of World War 2, the squadrons were sent to France as part of the BEF, their place being taken by 613 and 614 RAuxAF Squadrons. In 1942 the main E/W runway was extended to 5,100 feet/1,554 m, the SW/NE runway was extended to 4,200 feet/1,280 m and the SE/NW runway was abandoned. Army Co-operation activity continued with the more offensive Tomahawks and later Mustangs equipping the squadrons. With the approach of D-Day two additional T.2 hangars were erected, and the airfield was full to bursting point with aircraft and personnel supporting the landings in Normandy on 6 June 1944. As the Allies advanced in France, the Odiham squadrons were sent to forward airfields, although the station remained busy until the end of 1944, by which time Odiham was part of Fighter Command. Transport Command took over in June 1945, but the station was transferred to the RCAF on 21 October, 1945 still in the transport role with Dakotas until they disbanded.

Odiham was returned to Fighter Command on 28 June 1946 to become the first Vampire Fighter Wing in the RAF, with 247 Squadron responsible for the introduction of the type. Both 54

Below: Odiham-based 54 Squadron formed an aerobatic team with Hunter F.1s in 1955 from February to September, when they then re-equipped with the improved Hunter F.4. *Hawker*

Above: Odiham is now the home to the RAF fleet of Chinook helicopters, HC.2 ZA681 of 27 Squadron is seen in April 1996 with a "C" Type hangar and the top of the original watch tower in the background. *Author*

and 72 Squadrons were equipped during July 1946, although at the end of January 1947 130 Squadron was renumbered 72 Squadron, and the improved Vampire F.3s arrived in 1948. In July 54 Squadron became the first to fly jet fighters across the Atlantic while conducting a goodwill tour of Canada and the USA, the highlight being the New York Centenary celebrations. The Vampires left Odiham and flew to Goose Bay via Iceland and Greenland, arriving on 14 July. At the end of 1949, the Wing was re-equipped with Vampire FB.5s, bringing a ground-attack capability in addition to the interceptor role, and 72 Squadron left for North Weald in March 1950, to be replaced by 421 (RCAF) Squadron in January 1951, who trained on Vampires before returning to Canada to convert to Sabres. From 1951, 54 and 247 Squadrons converted to Meteor F.8s which they flew for the next four years, participating in Fighter Command and NATO exercises. To cope with the jet fighter operations the E/W runway was extended to the east to 6,000 feet/1,838 m. A further T.2 hangar was erected on the south side and new hardstandings constructed in front of the main technical site.

In the spring of 1954, the resident squadrons moved out temporarily while Odiham was prepared for the spectacular Review of the RAF by Her Majesty Queen Elizabeth II on 15 July. The static display consisted of 318 aircraft from Chipmunk trainers to Washington bombers, and the flypast, led by a solitary Sycamore helicopter, consisted of 641 aircraft in 47 formations, the last in line being the Supermarine Swift. With the station returned to normal, 46 Squadron was formed on 15 August 1954 with Meteor NF.12s and NF.14s to make up the night fighter element of the Wing. The following year 54 and 247 Squadrons received their long-awaited Hunters, initially the F.1s, then F.4s and in 1957, the more effective F.6s. Defence cuts resulted in the disbandment of 247 Squadron at the end of 1957.

The first unit to take delivery of the Gloster Javelin was 46 Squadron when the first aircraft arrived at Odiham on 24 February 1956, starting the replacement of the Meteors, and undertaking service flying trials. The introduction of the Javelin was somewhat challenging, as the aircraft still suffered development problems. These included restrictions on manoeuvres in the looping plane due to a risk of super stall. It was the first purpose-built all-weather fighter with the RAF, it was the world's first twin-jet delta fighter and at over 31,000lb it was by far the heaviest fighter to enter service with the RAF. The squadron personnel had an underlying feeling of pride in being selected to introduce this new combat type to the RAF, despite some trepidation. The first loss was when Wing Commander Birchfield DFC and his navigator crashed in a Javelin on the approach to Odiham in bad weather at night during June. The intensive flying programme required reaching 1,000 flying hours as quickly as possible, practising every operational task, the target being achieved within eight weeks. To achieve this task required pairs of Javelins to take-off at 30-minute intervals from 08.30 to 14.30 hrs, and again from 18.30 to 02.30 hrs the following morning covering the night exercises. During these trials the Gloster-designed maintenance features helped to achieve the flying targets. There was no dual-control Javelin initially and it took some six weeks to convert all the squadron aircrews due to the small number of aircraft available, but it was found easy to fly and robust. The Javelins of 46 Squadron continued to maintain night defence with FAW.1s to FAW.6s until it was decided to disperse the Wing and close Odiham as a fighter base. The Javelins of 46 Squadron left for Waterbeach and the Hunters of 54 Squadron left for Stradishall on 15 July 1959, the station then being reduced to C&M. Odiham then reopened under Transport Command in February 1960 as a helicopter base, and it is now the main station for RAF Chinook operations.

Stradishall, Suffolk;
TL720515, 7 miles NE of Haverhill, by A143.

As part of the 1930s RAF Expansion Scheme, Stradishall was built as a bomber station, but served in all home-based commands apart from Army Co-operation. There were initially five 'C' Type hangars on the technical site located on the north side of the landing ground, with brick-built 'H' barrack blocks, station HQ, workshops, stores and administration buildings. Later three T.2 hangars were also added. Construction was far from complete when the station opened on 3 February 1938 and IX Squadron with Heyfords and 148 Squadron with Wellesleys began to arrive on 10 March. As the station was to prepare for the arrival of Wellington bombers, it was one of the early bases to feature a concrete runway, when a 3,000ft strip was started late in 1938. This runway was later extended to 6,000 feet and two more were added with 36 heavy bomber-type dispersals. The station was initially used for bombing duties when 214 Squadron arrived with Wellingtons in February 1940, remaining at Stradishall until October 1942, by which time the squadron had converted to Stirlings, starting operations on 18 May 1942. From October 1942 1657 Conversion Unit replaced 214 Squadron to train crews on the Stirling bombers, which continued until disbandment on 15 December 1944. Two days later Stradishall became part of 3 Group Bomber Command and on 17 December 1944 186 Squadron arrived with Lancaster IIIs, remaining until disbandment on 17 July 1945. After a brief period in the transport role, bombers returned on 19 September 1946 with the Lancasters of 35 Squadron, joined by 115 Squadron on 27 September and 149 and 207 Squadrons during November 1946. The Lancasters departed for Mildenhall in February 1949, when Stradishall was reduced to C&M on 18 February.

On 27 April 1949 it was decided that Stradishall should become an advanced flying school (AFS) within Fighter Command, and it was reopened on 6 July 1949, with 226 OCU

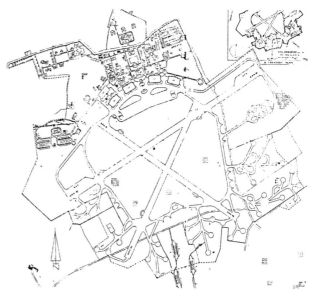

Above: In 1957 the main runway at Stradishall was extended at both ends and ORPs added by each threshold. The main technical site was on the northern boundary of the airfield with four 'C' Type hangars in a curved layout with a large ASP across the front. The remainder of the technical and accommodation site was located to the north of the hangars with the World War 2 dispersals retained around the perimeter track. The station closed to flying on 27 August 1970, and the site subsequently became a prison. *RAF Museum*

reforming on 1 September with Spitfires, Meteor F.4s and T.7s and Vampire FB.5s until January 1951. In March 1955 in preparation for the return to operational status, a large ASP was built in front of the hangars and ORPs were laid at each end of the main runway, with the dispersals on the west and southwest areas given blast wall protection. On 31 March 125 Squadron re-formed with Meteor NF.11s, which were replaced by Venom NF.3s in January 1956, but the squadron disbanded on 10 May 1957. From 9 June to 18 July 253 Squadron worked up with Venom NF.2s before moving to Waterbeach, and 89 Squadron re-formed at Stradishall with Venom night fighters from 15 December. In the second half of 1956 152 Squadron with Meteor NF.12/14s was based at Stradishall and returned with 263 Squadron on 18 August 1957, remaining until 8 July 1958. The longest-staying resident at the time was 89 Squadron, which converted to Javelin FAW.2s and FAW.6s in October 1957, and they were

Left: Stradishall-based Javelin T.3 XH435 Z:85 Squadron was used for continuation training and the squadron was based at Stradishall with Javelins from 30 November 1958 until moving to West Malling on 5 June 1959. *J. J. Halley*

joined by 85 Squadron with Javelins from 30 November 1958 to 7 September 1959. By this time the station had become a short-range day fighter base with 1 Squadron Hunter F.6s arriving on 2 July 1958, and joined by 54 Squadron on 1 August. Both squadrons converted to Hunter FGA.9s before 1 Squadron left for Waterbeach on 7 November 1961, followed by 54 Squadron on 21 November. Having been part of 12 Group Fighter Command since the autumn of 1959, in 1960 208 Squadron worked up with Hunter FGA.9s from 1 April until 30 May, ready for an overseas posting. Amongst other units using Stradishall were 111 Squadron from 15 July to 13 September and 43 Squadron between 23 September and 13 October 1960. Stradishall transferred to Training Command on 1 December 1961 as the home of 1 Air Navigation School (ANS), which finally departed on 27 August 1970 when the station closed. The technical site has been completely cleared on the airfield side of the camp and replaced by Highpoint Prison. All sign of the runways and dispersals have been removed and the area returned to agriculture.

Tangmere, W. Sussex;
SU910060 3 miles E of Chichester, by A27.

For most of its operational life, RAF Tangmere was a fighter station, the original site being requisitioned on 25 September 1917 and the technical site to the north of the landing ground included three double Belfast hangars and one single unit between a pair of the doubles. Station HQ, stores, workshops, MT yard, administration buildings were located behind the hangars alongside the A27 road, with accommodation blocks on the other side of the road. In addition a line of large hangars were constructed to house Handley Page O/400 bombers with the US Army Air Service as a Training Depot Station (TDS). Flying operations started with 92 Squadron SE.5as from March 1918, the Americans finally taking over in September, the hangars being virtually complete. With the signing of the armistice, the residents returned home across the Atlantic, leaving the airfield in use for training until it was closed in 1920. However, the base was retained by the Air Ministry and reopened on 1 June 1925 as a storage site using the extensive hangars. The station HQ was re-formed on 23 November 1926 and the airfield re-activated with 43 Squadron (The Fighting Cocks) moving in with Gamecocks in December, to be joined by 1 Squadron on 1 April 1927, being fully equipped with Siskin IIIas by November, starting the fighter role of the station.

From mid-1928 the earlier wooden accommodations were replaced by more permanent brick-built messes, barrack blocks and married quarters, which were completed in 1930 with a grass landing ground to the south with a 3,600 feet NE/SW landing run. The Fighting Cocks received the Hawker Furies from May 1931, followed by 1 Squadron in February the following year, the two squadrons remaining in residence until the re-equipment with Hurricanes from October 1938. The station soon went on a war footing, with buildings camouflaged and defences prepared. For the Battle of Britain, Tangmere became a sector station within 11 Group Fighter Command, supporting the BEF in France with a satellite at Westhampnett, the squadrons being responsible for the defence of the docks and installations along the south coast including Portsmouth. Two concrete runways were laid, with dispersals off a connecting perimeter track. On 16 August 1940 an attack by enemy bombers left the station seriously damaged, with two hangars completely destroyed, the other three badly damaged and a number of other buildings including the officers' mess wrecked. Ten servicemen and three civilians

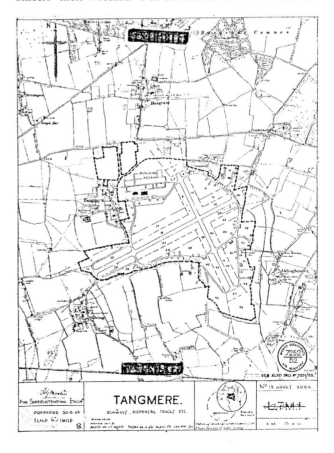

Right: In 1952 both runways at Tangmere had been extended for jet fighter operations, the main runway extension being to the west and the cross runway to the southeast. The original Belfast hangars were destroyed by enemy action and had been replaced with T.2 hangars on the north-side-located technical site, and the base's main contribution to the early part of the Cold War was to be home to 1 and 43 Squadrons. For the Hunter operations ORPs were constructed at both ends of the main runway and an ASP along the front of the three T.2 Hangars. The main part of the camp was located behind the hangars to the north of the airfield. The station closed to flying on 16 October 1970 and now only the T.2 hangars remain for storage. *RAF Museum*

Above: The traditional 1 Fighter Squadron group taken at Tangmere in October 1954, with the Meteor NF.11s of 29 Squadron in the background. One of the T.2 hangars is to the right. *Eric Snowdon collection*

were killed, with another 20 injured and a number of aircraft destroyed. The airfield remained operational throughout, and as a result of the raid aircraft were dispersed and protected by blast pens. Following the Battle of Britain, the station went on to the offensive with intruder sweeps over occupied France, with both 1 and 43 Squadrons returning from time to time. Tangmere became a base for Typhoons on ground attack operations across Europe as part of the 2nd TAF and the airfield was taken over by six RCAF squadrons in support of the D-Day landings, the squadrons progressively moving out as advances were made on the Continent, after which the station became relatively quiet.

In late 1945 a refurbishment programme commenced with three T.2 hangars erected on the sites of the destroyed Belfast hangars and the main runway extended to the SW. The only flying residents were 85 Squadron with Mosquito NF.XXXs, later replaced by NF.36s, and they were joined by the returning 1 Squadron on 30 April 1946 with Spitfire F.21s, to be replaced by Meteor F.3s in October. In the same month 222 Squadron arrived with Meteor F.3s to complete the Tangmere Wing, 85 Squadron providing the night defence until April 1947, when it was replaced by 266 Squadron also operating Meteor F.3s. In June 1948, 1 Squadron re-equipped with the newer Meteor F.4s, and on 11 February 1949, 222 Squadron was renumbered 43 Squadron, bringing together again the traditional rivals. The standard Fighter Command station establishment changed to one day and one night fighter squadron, and the Fighting Cocks moved to Leuchars in November 1950 to be replaced by Mosquito NF.36 equipped 29 Squadron from West Malling.

The Mosquitos were replaced by Meteor NF.11s in August 1951, Tangmere being an all Meteor base for the next four years.

Meanwhile the RAF High Speed Flight was formed on 14 June 1946 to retain the Absolute World Speed Record for Britain. The course was set up along the south coast with Group Captain E. M. (Teddy) Donaldson, and Sqdn Ldrs W. A. (Bill) Waterton and Neville Duke to fly modified Meteor F.4s for the record attempt. The two modified Meteors were delivered in August and on 7 September Donaldson achieved 616 mph/991 km/h, successfully retaining the record for Britain. Teddy Donaldson later became the air correspondent of The Daily Telegraph, Bill Warterton the chief test pilot of Gloster with a responsibility for the early Javelin development testing, and Neville Duke the chief test pilot of Hawker Aircraft, leading the flight testing of the Hunter. Neville Duke returned to Tangmere in August 1953 with the red painted modified Hunter prototype WB188 to further raise the speed record, and in perfect conditions on 7 September raised the Absolute Air Speed Record over the same course to 727.63 mph/1,171 km/h, equivalent to Mach 0.92 at sea level.

My good friend Eric Snowdon was a Cold War warrior with 1 Squadron at Tangmere flying both Meteors and Hunters. The RAF had drastically reduced its personnel after the end of World

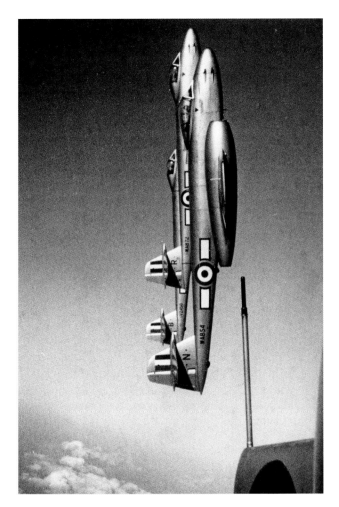

Left: A flight of Meteor F.8s of 1 Squadron in a vertical loop from Tangmere in November 1953. No 1 Squadron operated Meteor F.8s from August 1950 until replaced by Hunter F.5s from September 1955. *Eric Snowdon collection*

War 2, and with the threat of the Cold War, began recruiting air crew to fly the new jet fighters coming into service. Eric joined the RAF for an eight-year short service commission in 1950 and had his basic flying training on Chipmunks at No.1 Basic Flying Training School at Booker. He then moved to 6 FTS at Tern Hill where his intermediate training was on Harvards before moving to 207 AFS at Full Sutton in Yorkshire to fly Meteor F.4s. Eric completed his flying training with 226 OCU at Stradishall flying Meteor F.8s fitted with the early cockpit canopies with a metal fairing at the rear.

Eric then joined 1 Squadron at Tangmere in mid-1953 which was equipped with Meteor F.8s and he lived in the brick-built officers' mess within the main technical site. For the average sortie the endurance of the Meteor F.8 was 50 to 60 minutes on internal fuel including the ventral tank, and the plan was to land with 40 gallons a side, otherwise if forced to overshoot there may not be enough fuel for another circuit. The training continued on the squadron with air-to-air gun firing on targets towed by another Meteor, practice dog-fighting, tail chases and quarter attacks on other fighters or sometimes Lincoln bombers. Battle formation was practised with two pairs of aircraft spread out to ensure that the number twos could protect the rears of the number one. The gyro gun sight had a cine camera recording the performance of the target acquisition, and when USAF B-36 bombers were used as practice targets, the wingspan was too great for the graticules in the gun sight to register, so the tailplane was used instead.

Navigation at low levels using map reading was also practised, usually at around 500 feet, and regular position reports were made over the radio using grid references. Long range navigation exercises were practised usually flying at 45,000 feet, cruising at 350 to 400 knots which increased the endurance to up to 90 minutes, allowing a triangular route from Tangmere to Liverpool, Newcastle and return to Tangmere as an exercise. A number of pilots were killed in the Meteors when turning downwind to land. With air brakes out, the turn was made with one-third flaps selected and the undercarriage lowered. However, the starboard main wheel usually came down first, causing the wing to stall and drop, resulting in loss of control and a crash. When the cause was found, the answer was to retract the airbrakes before selecting flaps and undercarriage.

In mid-1955 1 Squadron converted to the Sapphire-powered Hunter F.2s and F.5s which were more reliable than the Avon-powered Hunters at that time. The Hunter was supersonic in a shallow dive and Eric managed to reach 58,000 feet, but only just. The Hunters had nearly one hour of endurance on internal fuel and the operational training continued much as before, but with better performance. Eric completed his service with 1 Squadron as a flight lieutenant and deputy flight commander before posting to 233 OCU at Pembrey where he completed his flying career with the RAF training Hunter pilots.

Meanwhile ORPs were constructed on both ends of the runway and an ASP located along the front of the hangars. On 1 August 1954 34 Squadron re-formed with Meteors and eighteen months later both 1 and 34 Squadron received the Sapphire-powered Hunter F.5s. In August 1956 both squadrons went to Cyprus for the Suez Campaign, returning in December. On 14 January 1957 29 Squadron moved to Acklington and they were not replaced for nine months until 25 Squadron arrived with a mix of Meteor NF.12/14s. With the decision to concentrate air defence on the V-Bomber bases, Tangmere was in the wrong location, and during 1958 all three squadrons were disbanded, with the station handed over to Signals Command. Flying reduced drastically from 1 October 1963 and the station closed on 16 October 1970. Many of the buildings were demolished from February 1963, but the three T.2 hangars were retained for grain storage, and the site is now an industrial and housing estate. It is still very much worth a visit as the Tangmere Aerodrome Memorial Museum is located on the boundary of the old airfield and amongst the fine exhibits are the record-breaking Meteor F.4 and Hunter Prototype.

Thornaby, North Yorkshire;
NZ455163, 2 miles SW Middlesborough by A1045.

RAF Thornaby played a modest part in the Cold War with the reformation of 608 RAuxAF Squadron in July 1946 equipped with Mosquito NF.30s, replaced in 1948 by Spitfire F.22s. In 1950 jets replaced the Spitfires with the arrival of Vampire F.3s, followed by FB.5s in 1952. Soon after the re-equipping with Vampire FB.9s, the squadron disbanded with all the other RAuxAF units in March 1957. The only regular RAF fighter unit to be based at Thornaby was 92 Squadron with Hunter F.6s which arrived from Middleton St George on 30 September 1957, and returned to its original base on 1 October 1958, after which the airfield closed.

Thornaby was used as a landing ground during World War 1 and the formal airfield was built in 1928, coming under the control of RAF Coastal Command in 1939. The technical site on the NW side included two 'C' Type hangars and a Bellman hangar with a selection of pre-Expansion Period buildings. Initially there were three grass runways, but these were paved in 1942. Little now remains of the station, with residential development over the site, but a memorial and replica Spitfire have been erected on the A1045 roundabout on the site boundary.

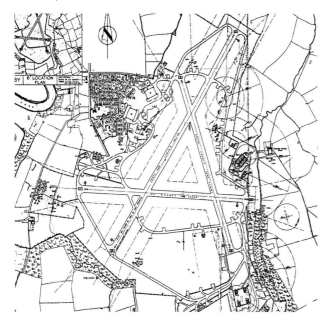

Above: By 1945 Thornaby had the main runway extended in both directions beyond the World War 2 boundaries with the other two runways extended from one end. The main technical site consisted of two 'C' Type hangars with the remainder of the technical and accommodation buildings in the northwest corner. In addition to being the home of 608 Squadron RAuxAF until it disbanded in March 1957, the Hunter-equipped 92 Squadron was the only postwar regular RAF unit. The station closed on 1 October 1958 and has been fully built over. *RAF Museum*

Waterbeach, Cambs;
TL495665, 6 miles NE of Cambridge by A10.

Waterbeach is where 56 Squadron attempted unsuccessfully to introduce the Supermarine Swift to RAF service. The Swift had a greater range than the similarly R-R Avon-powered Hawker Hunter, but the Hunter also suffered problems in its development with a short endurance in the early versions, and problems with the Avon engine stalling when the guns were fired. Although the Swift was withdrawn as a day fighter, it gained considerably more success in the -reconnaissance role.

Construction of the airfield at Waterbeach started during the winter of 1940-41 to the more spartan specification than used in the earlier Expansion Scheme. Intended as a bomber station, two 'J' Type hangars were erected in a semicircle with a gap between them as a provision for a third which was never built. The technical site was located in the SE corner and three runways were constructed, connected by a perimeter track with aircraft dispersals. The bomb dump was located on the northern boundary of the landing ground. The Station HQ was opened on 11 January 1941, and the first arrivals were the Wellington bombers of 99 Squadron on 19 March, flying from the station until their last operation to Emden on 14 January 1942. In January 1942, 1651 Conversion Unit was formed to train crews on the Short Stirling heavy bomber until November 1943, and a return was made to the operational role when 514 Squadron arrived on 23 November with Lancaster IIs, their first operation being on 15 November 1943. No. 514 Squadron continued on bomber operations until their final raid on 24 April 1945, and the squadron disbanded on 22 August 1945.

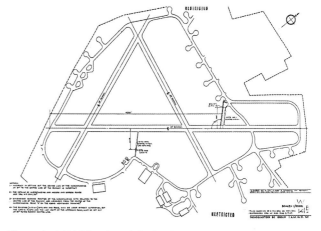

Above: In 1954 Waterbeach had a three-runway layout with the main runway extended to the northeast. The rather basic technical site was in the southeast corner with two 'J' Type hangars and a group of T.2 hangars nearby. The administration and accommodation, not shown on the plan, was also in the southeast corner. The original World War 2 weapons storage area was on the extended northern boundary. *RAF Museum*

Left: No 56 Squadron was tasked with introducing the Swift into RAF service from August 1954 to March 1955, but without success. Swift F.3 WK240/7300M E:56 Squadron was relegated to ground instruction duties after the type was withdrawn and temporarily replaced by Meteor F.8s until Hunters were available. *J. J. Halley*

Waterbeach then changed to Transport Command with RAF-operated Liberator transports of 59 and 220 Squadrons with services to Europe and Asia, these duties continuing until December 1947, to be replaced by Dakotas of 18, 53, 62 and 77 Squadrons from 8 December, which stayed until February 1950. Waterbeach then was allocated to Fighter Command on 1 March 1950 with an extension to the runway including QRAs, and ORPs for each squadron. Jet fuel installations were supplied and a weapons store was built, ready for the arrival of the Meteor F.4s of 56 and 63 Squadrons. In December 1950, with increasing tension from the Soviet Union, the squadrons were ordered to participate in the top secret exercise Fabulous in December 1950. With a high expectation of an unprovoked Russian attack, each squadron would have aircraft fully armed and fuelled at readiness. Standby pilots were sitting in the cockpits awaiting orders and others would be at readiness in full kit awaiting a call to scramble. Specifically, there was to be no delay to prevent a rapid departure in the event of the order coming from the Eastern Sector Operations Room at Neatishead. This high level of readiness had never before been ordered in peacetime. Waterbeach continued to maintain a high level of readiness for many years, the QRA aircraft being protected by concrete walls built on the dispersals from 1953-54. The resident squadrons were re-equipped with Meteor F.8s late in December 1950, and then 56 Squadron received its first Swift F.1 in February 1954, the first swept-wing fighter for Fighter Command. In addition to being inadequately armed and underpowered, the aircraft was difficult to handle with a

No 253 Squadron operated Venom NF.2As from Waterbeach from 18 April 1955 until 31 August 1957, followed by disbandment on 2 September. *J. D. R. Rawlings*

tendency to tighten in high-speed turns from which recovery was difficult. Following a number of fatal accidents, the Swift was withdrawn and 56 Squadron returned to Meteor F.8s in March 1955, until May when the new Sapphire Hunter F.5s were introduced. In November 1956, 63 Squadron exchanged their Meteors for Avon-powered Hunter F.6s. In addition to the original two 'J' Type hangars, three T.2s and a B1 hangar were also constructed on the technical site.

Indecision by the Government led to delays in the development of a suitable modern night fighter to replace the Mosquitos, which was to result in the Gloster Javelin. The adaptation of the Meteor to the night fighter role was not a total success, and as an interim, de Havilland developed the Venom Night Fighter following the earlier Vampire NF.10. The overall policy was to have two day fighter squadrons providing the readiness, with an all-weather fighter squadron added to form a wing. At Waterbeach the night fighter element was provided by 253 Squadron which re-formed in April 1955 with Venom Night Fighters, but as part of the reduction in Fighter Command, the squadron was disbanded in August 1957. They were replaced by 153 Squadron in September with Meteor Night Fighters and in June 1958 they were renumbered 25 Squadron. Then 63 Squadron disbanded on 24 October 1958, with 56 Squadron departing in July 1959, immediately being replaced by the Javelin FAW.2s of 46 Squadron, which remained until May 1961. No. 25 Squadron had their Meteors replaced by Javelin FAW.7s in March 1959, and at the end of the year with the ultimate FAW.9, leaving Waterbeach in November 1961. The Javelin FAW.9s of 64 Squadron replaced 46 Squadron from 27 July 1961 to 13 July 1962. Meanwhile the crews of 60 Squadron trained at Waterbeach in early 1961 before departing for Singapore on 25 June.

With a further reduction in Fighter Command, Waterbeach was taken over by 38 Group, Transport Command in January 1962 with the Hunter FGA.9s of 1 and 54 Squadrons with the duties of ground attack and escort of air transports. The Hunters had a short stay and left for West Raynham on 8 August 1963, and the airfield was used for 'touch & goes' by the Varsities and Jetstreams from 5 FTS at nearby Oakington before the RAF station closed. It was taken over by the Army as Waterbeach Barracks, which it still is today, and the main runway still exists, as do many of the original RAF buildings including some of the hangars. As a reminder of past glories, Hunter F.2 WN904 is displayed on the gate in 1 Squadron markings.

Wattisham, Suffolk;
TM025510, 6 miles SW of Needham Market, by B1078.

As part of the RAF Expansion Scheme, RAF Wattisham opened in March 1939 with four 'C' Type hangars fronting the traditional technical site on the east side of the landing ground. The first arrivals were the Blenheims of 107 Squadron at the end of April 1939, to be joined by 110 Squadron on 11 May. Theses two Squadrons formed 83 Wing as part of the Second Echelon of the Advanced Air Striking Force, going into action in May 1940, when they sustained heavy

No 41 Squadron was re-formed with Javelin FAW.4s on 1 February 1958 and moved to Wattisham on 5 July 1958. The squadron operated Javelin FAW.4s until February 1960 and was disbanded on 31 December 1963. Javelin FAW.4 XA767 K:41 Squadron was on the Wattisham ASP with other Javelins of the squadron being prepared for flight including refuelling.
RAF Museum P.019630

Above: No 152 Squadron operated Meteor NF.14s at Wattisham from 30 June 1954 when it was re-formed, until July 1958. The squadron was lined up on the ASP in 1955 with The Wattisham 'C' Type hangars in the background. *RAF Museum P.019571*

losses on anti-shipping operations, which were probably the one of the most hazardous type of operations of the entire war. The station itself also suffered damage from enemy bombing attacks, including the destruction of a barrack block and a hangar set on fire. With the departure of the Blenheims in August 1942, Beaufighters were operating on maritime patrols until September when the station was allocated to the Americans. In preparation for the new occupants, three concrete runways were laid and taxiways, but instead of the anticipated bombers, the base was used for the maintenance of USAAF fighters based in Britain. Wattisham became operational again on 15 May 1944 when the P-38J Lightning-equipped 479th FG arrived as part of the 8th AF on bomber escort and ground-attack duties, including support of the D-Day landings in Normandy. In September 1944, the Lightnings were replaced by P-51D Mustangs which continued with similar duties to the end of the war, being credited with the last enemy aircraft shot down by the 8th AF on 24 April 1945.

With the departure of the Americans, Wattisham returned to RAF control and was placed under C&M until August 1946 when Fighter Command took control. On 4 November 1946 266 Squadron arrived with Meteor F.IIIs to be joined by 56 Squadron in mid December. However, they only stayed until mid-April 1947, as the runways were considered inadequate for the Meteors, and in October the construction of a 6,000ft main runway was approved, the station being designated as an interceptor fighter base. With the new runway completed on schedule by September 1950, 257 and 263 Squadrons arrived from Horsham St Faith, both units with Meteor F.4s, which

Above: The technical site at Wattisham was on the eastern side of the airfield with four 'C' Type hangars in a gentle semi-circle and a large ASP in front. The station administration and workshops were behind the hangars with the accommodation across a small road including the traditional airmen's 'H' barrack blocks.
RAF Museum

Above: No 263 Squadron moved to Wattisham on 22 November 1950 and operated Sapphire-powered Hunter F.5s from May 1955 to August 1956 as well as Hunter F.2s from February 1955. The squadron deployed to Wymeswold with 257 Squadron from 10 June 1956 until 15 January 1957. A mixture of Hunter F.2s and F.5s were on the ASP at Wattisham with 'C' Type hangar No 1 in the background and the early style air traffic control. *RAF Museum P.019630*

UK Airfields of the Cold War   55

were soon replaced with the more effective Meteor F.8s. Initially each squadron was equipped with 16 aircraft, which in late 1951 was further increased to 22 aircraft each. The station was also to receive a night fighter capability and as a result 152 Squadron was formed on 30 June 1954 with a mixture of Meteor NF.12s and NF.14s. Further improvements were made to the main runway and plans were made for walled revetments to protect the fighters against blast damage. A large ASP was constructed and ORPs were constructed on the south side of the main runway at both ends.

In November 1954, 257 Squadron became the first to be equipped with the Sapphire-powered Hunter F.2s, 263 Squadron receiving theirs the following year. In August 1956, the Sapphire Hunters were replaced by Avon-powered Hunter F.6s. With the reduction in the fighter forces in 1957, 257 Squadron was disbanded on 31 March, while 152 Squadron departed to Stradishall on 28 August, followed by 263 Squadron the next day, leaving Wattisham to be prepared for the Lightnings. Before the arrival of the new Mach 2 interceptor, Wattisham was used by 111 Squadron 'Black Arrows' aerobatic team, which arrived on 18 June 1958, and including a formation loop with 22 Hunters in its programme. At the end of June 1958 41 Squadron arrived with Javelin FAW.4s and 5s, and in 1960 became the first to operate the Javelin FAW.8s. On 10 July 1959 56 Squadron arrived with Hunter F.6s from Waterbeach in preparation for re-equipping with Lightning F.1as in January 1961 followed by 111 Squadron in April, heralding a new era for Wattisham.

West Malling, Kent;
TQ680555, 5 miles W of Maidstone by B228

West Malling was so badly damaged by the Luftwaffe in the summer of 1940 that it was inactive during the majority of the Battle of Britain, later becoming one of the RAF major all-weather fighter stations. After use as a small landing ground in World War 1, the site was returned to aviation in 1930 as a private aerodrome, in 1932 becoming Maidstone Airport and was requisitioned by the Air Ministry at the start of World War 2. Improvements and extensions were made with the first RAF arrivals being 26 (Army Co-operation) Squadron in June 1940 with Lysanders. The station then became a major part of the anti-invasion home defence and was one of the RAF Fighter Command airfields in 11 Group, as a satellite for Kenley and an advanced airfield for both Kenley and Biggin Hill. The first squadron to arrive was the Defiant-equipped 141 Squadron on 12 July 1940, but after losing half the squadron to the Luftwaffe on 19 July, the Defiants were withdrawn, leaving 26 Squadron as the sole occupants. The first enemy raid was on 10 August which caused little damage, but the next one on

Below: No 85 Squadron moved to West Malling on 5 June 1959 and operated Javelin FAW.8s from March 1960. The squadron moved to West Raynham on 6 September 1960, continuing to fly FAW.8s until March 1963. Javelin FAW.8 XH992 was resident at West Malling in August 1959. *RAF Museum P.017971*

Left: No 85 Squadron operated Meteor NF.14s from April 1954 until November 1958. The squadron was West Malling-based until a move to Church Fenton on 18 September 1957 with disbandment on 30 November 1958. Meteor NF.14 WS775 of 85 Squadron ready to join the West Malling circuit. *RAF Museum P.019061*

Below: No 25 Squadron moved from Germany to West Malling on 14 May 1948 with Mosquito NF.36s as part of the night defence of Britain. The Mosquitos were replaced by Vampire NF.10s in July 1951 as an interim night fighter until Meteor NF.12s became available in March 1954. The Vampire NF.10 is plugged in to the trolley-ack battery starter. *de Havilland*

15 August caused considerable damage and killed two airmen. The next day a high-level attack damaged three Lysanders, followed two days later by a dive-bombing attack which destroyed three more Lysanders and caused further damage to the site. There were three more attacks during August and a further one on 3 September, when due to wreckage and rubble 26 Squadron finally moved out to Gatwick. By 12 September there were signs of the station recovering, but a lone raider dropped anti-personnel bombs, killing six soldiers. Before the end of September there had been five more raids on West Malling. Although bombing continued throughout October, sufficient repairs were made to allow 66 Squadron with Spitfires to arrive on 30 September and remain until 24 February 1941 when they moved to Exeter. This allowed repairs to continue and to prepare the station for the night defence of London and the southeast against enemy bombers. The accommodation was rather basic, initially with two Sommerfield track runways, one 'J' Type hangar and sixteen Blister hangars dispersed around the perimeter. The personnel were accommodated in local requisitioned properties.

The first night fighters to move in on 14 April 1941 were 264 Squadron with Defiants again, but they were more vulnerable in night operations, and were joined by 29 Squadron flying the more effective Beaufighter Ifs on 27 April. The night fighting role continued with 29 Squadron being replaced by Mosquito night fighters of 85 Squadron on 13 May 1943 commanded by Wg Cdr John Cunningham, the World War 2 night fighter ace, the squadron remaining resident for an extended period. Because of its closeness to occupied Europe, West Malling was used frequently as a forward base by intruder squadrons. With the V-1 flying bomb offensive in June 1944, West Malling became an important base to combat these fast-flying Doodlebugs, as they were known, to catch them over the Channel before they were in the range of the ground-based anti-aircraft guns. Included in the anti-diver patrols against the V-1s were 85 Squadron with Mosquito NF.XIX, joined by 157 Squadron on 21 July until 29 August when the main threat had passed. The station then closed for major reconstruction with a 6,000ft concrete runway replacing the longer of the two metal tracking runways, the upgrade being completed by June 1945.

On 29 October 29 Squadron returned with Mosquito NF.XXXs, replacing them with the ultimate NF.36s in July 1946, the squadron rotating out to Lubeck until leaving for Tangmere on 25 November 1950.

The night fighter strength grew with the arrival of the reformed 500 (County of Kent) RAuxAF Squadron on 10 May 1946, receiving Mosquito NF.XIX in February 1947. These were replaced by Mosquito NF.30s from April and the squadron changed to the day fighter role in May 1948 when Spitfire F.22s arrived. The squadron entered the jet age in July 1948 when it was re-equipped with Meteor F.3s, to be followed by F.4s and finally F.8s before disbandment on 10 March 1957 with all the other RAuxAF squadrons. Meanwhile the regular RAF was represented by 25 Squadron with Mosquito NF.36s from September 1946 and on 16 April 1947 85 Squadron returned with Mosquito NF.36s. However, with the heavy all-up weight of the ultimate Mosquito night fighter, it was no match for defence against a potential attack by Soviet bombers, and as an interim measure, Meteor night fighters arrived at the Station. On 28 February 1955, 153 Squadron re-formed with Meteor NF.12 and NF.14s, leaving for Waterbeach on 17 September 1957. With the strategic situation changing, ballistic missiles were beginning to replace manned bombers, and although West Malling was in the front line, it was no longer effective. On 5 June 1958 85 Squadron returned briefly to West Malling with Javelin FAW.6s and later FAW.8s, but moved out to West Raynham on 6 September 1960, finishing the station as an operational base. It was placed under C&M until 1970 when Kent County Council bought the site, and although it was used by commercial aviation for a while, it has now been fully developed for commercial purposes with industrial and residential development covering all trace of the previous rather austere airfield.

West Raynham, Norfolk;
TF850245, 2 miles from village, off A1065

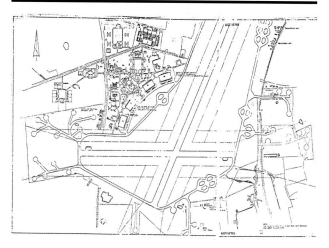

Above: In 1955 West Raynham had a two runway layout with the main runway extended to the north for operations by jet fighters. The technical site on the northwest side consisted of four 'C' Type hangars in a gentle semi-circle and later had an ASP built across the front of the hangars. Behind the hangars was the main administration and support accommodation with living accommodation across a local road, including four 'H' blocks. *RAF Museum*

Once the centre of fighter excellence, this remote Norfolk station was the home of the Central Fighter Establishment (CFE), where the latest fighters to enter service with the RAF were operated to develop the most effective tactics for use in service. From what was developed here by the experienced pilots, the squadron pilots could progress from learning how to fly the aircraft, to learning how to make the best use of them in combat. Built to the standard Expansion Scheme layout, the station opened in 1939 as a bomber base with four 'C' Type hangars, two 6,000ft concrete runways and 36 heavy bomber dispersals located off a connecting perimeter track. The main technical site was on the NW side of the airfield and the watch tower was located in the centre of the hangar curve with the technical, administration and accommodation to the rear.

Left: The Tri-National Kestrel Evaluation Squadron operated nine Kestrels from West Raynham from October 1964 until November 1965, a programme which lead both the RAF and US Marines to adopt the concept in the form of the developed Harrier. *HSA*

Above: No 1 Squadron with Hunter FGA.9s within 38 Group, operated from West Raynham from 13 August 1963 until moving to Wittering on 18 July 1969 as the first RAF squadron to convert to Harriers. Hunter FGA.9 XE624 B:1 Squadron fitted with underwing fuel tanks has started its take-off run on 4 September 1968. *Author*

The first bombers arrived with 139 Squadron on 30 May 1940 were Blenheim IVs, moving out to Horsham St Faith the following month. They were replaced by 101 Squadron, who began operational flying on 4 July 1940 also with Blenheim IVs. The Blenheim operations continued until May 1943, when the airfield was closed for the construction of the runways which were completed in November. On 1 December 1943, the station was transferred from 2 Group Bomber Command to 100 Bomber Support Group which brought in 141 Squadron on 4 December and 239 Squadron on 10 December 1943, both equipped with intruder Mosquito F.IIs, a task which continued until the end of the war.

The CFE arrived between October and December 1945 from Tangmere, with a wide assortment of aircraft. At West Raynham the concept of telescramble was developed with the aircraft at readiness on the ORPs at each end of the main runway which were laid-in the mid 1950s. At the same time large ASPs were constructed in front of the hangars with a new control tower nearer the centre of the airfield. The airborne combat flight technique, known as Snake, was devised to ensure that defending fighters were gathered rapidly in each sector for massed attacks on approaching Soviet bomber assaults well before they had approached the British coast. Within the CFE were a number of separate units specialising in a number of tasks. The Air Fighting Development Squadron (AFDS) was responsible for the tactical development of the jet fighters, including the Meteor day and night fighters, Vampires, Venoms, Hunters and finally the Lightning, although the latter was flown from Coltishall. The Day Fighter Leaders School (DFLS) initially flew Meteor F.4s and F.8s, and in March 1958 became the Day Fighter Combat School, and later the Fighter Leader School flying Hunters until leaving for Binbrook in the autumn of 1962. The Night Fighter Leaders School (NFLS) initially flew Mosquito night fighters, but later they were replaced by Meteor NF11s and NF.14s. In July 1950 it became the All Weather FLS which was equipped with Javelins under the reserve title of 176 Squadron. From 1959 the CFE began to reduce in size, absorbing the Fighter Weapons School, and the AFDS and AWDS departed for Coltishall, leaving the Hunters of DFCS and AWFCS at Raynham which closed down in October 1962.

West Raynham resumed an operational role in August 1960 when the Javelin FAW.8s of 85 Squadron arrived, remaining here until disbandment on 31 March 1963. The station then transferred to 38 Group when Waterbeach closed in August 1963 and the Hunter FGA.9s of 1 and 54 Squadrons arrived. On 18 July 1969 1 Squadron left for Wittering to convert to Harriers and 54 Squadron went to Coningsby in September 1969 to convert to Phantoms. Meanwhile, on the eastern boundary of the airfield were Bloodhound ground-to-air missile sites, and between October 1964 and November 1965 the Tripartite Kestrel Evaluation Squadron flown by RAF, German and USAAF pilots evaluated the concept of what was to lead to the Harrier. When the Hunters of 38 Group departed, two Canberra target facilities arrived, 85 Squadron in January 1972 and 100 Squadron re-formed on 1 February. On 19 December 1975 85 Squadron disbanded and 100 Squadron moved to Marham on 5 January 1976. On 1 August 1972 45 Squadron had re-formed with Hunter FGA.9s to provide operational training for 38 Group, departing for Wittering on 29 September 1972, West Raynham closing to flying. It remained active as a Bloodhound site until the missiles were withdrawn and the site closed on 1 July 1994, although it still remains in the ownership of the MOD and most of the station survives with part of the domestic site sold for housing. The runways have also now been removed.

Wymeswold, Leics;
SK585225, 3 miles NE Loughborough off A60.

The rather bleak and basic airfield at Wymeswold only played a small part in the Cold War with three regular RAF fighter squadrons based there for a short while. The airfield was built to the standard World War 2 pattern with three paved runways connected by a perimeter track, and a basic technical site on the eastern side including a pair of T.2 hangars. The station opened on 16 May 1942 as part of 7 Group Bomber Command and was used for the training of aircrew for Bomber Command and from October 1944, the training of Transport Command crews. This activity continued until 10 December 1947 when the station closed. In February 1949, Wymeswold reopened and on 3 May 504 (County of Nottingham) RAuxAF Squadron arrived with Meteor F.4s, which were replaced by Meteor F.8s in March 1952 until the closing down of the RAuxAF in March 1957. In 1950 the station came under the control of 12 Group Fighter Command and from August 1955 until 1957 two regular RAF squadrons used the station for short periods, the main runway having QRAs constructed at both ends. They were 257 Squadron with Hunter F.2s from 10 June 1956 until 15 January 1957 and 263 Squadron with Hunter F.5s from 10 June 1956, leaving for Wattisham also on 15 January 1957, after which the station closed. Much of the airfield survives with the runways, and the technical site used by industry. The T.2 hangars, gutted

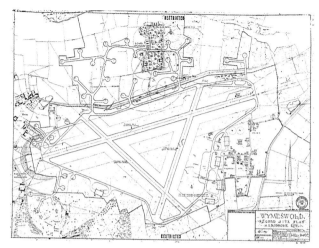

Above: Wymeswold in 1948 had a traditional three runway layout with the basic technical site including four T.2 hangars in the southeast corner of the airfield. There was an additional T.2 hangar on the southwest side which has since been removed. The main changes for Cold War jet fighter operations was to provide curved loops at the north side of the main runway across the earlier dispersals. Much of the airfield remains with the technical site now a commercial park. *RAF Museum*

control tower and water tower of this spartan station still exist, with a busy modern business park on the technical site. The main runway and perimeter tracks are used for motor sport, making this an overall busy and active location.

There were some 22 RAF stations used to varying degrees during the early stages of the Cold War, most of which are no longer in use by the RAF. With only basic changes, such as QRAs at the ends of the main runways and in some cases large ORPs to accommodate the first- and second-generation jet fighters, these airfields supported Britain's defences against Soviet aggression, ready to attempt to stop a massive aerial armada before nuclear weapons could be released. This was an extremely daunting task and required a high state of readiness for the young men of Fighter Command, much in the same way as the Battle of Britain, but waiting on standby instead of in constant action.

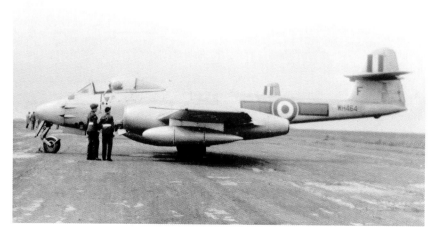

Left: No 504 Squadron flew Meteor F.8s at Wymeswold from March 1952 until disbanded on 10 March 1957. Meteor F.8 WH964 F:504 Squadron was at Wymeswold in June 1953. *RAF Museum P.019043*

Chapter 4

RAF Tactical Bomber Stations

With the end of World War 2, the main heavy bomber with the RAF was the Lincoln, derived from the wartime Lancaster. A total of 23 squadrons of Lincoln B.2s were operational within Bomber Command, the first replacing Lancasters from January 1946, and the last being replaced by Canberras in December 1955. From June 1950, these rather antiquated long-range bombers were supplemented by eight squadrons of USAF-surplus B-29-derived Washingtons. These aircraft operated mainly from Coningsby and Marham, and were capable of carrying an atomic bomb, with replacement by Canberras from early 1953, the last Washington being withdrawn in May 1954. Britain's first jet bomber was the Canberra, which was unarmed and capable of higher speeds and altitude, with the comfort of a pressurised cabin for the crew of three, consisting of a pilot and two navigators. The first Canberra, WD936, was delivered to 101 Squadron at Binbrook on 25 May 1951, there eventually being 23 Squadrons of Canberras in Bomber Command, the last being 35 Squadron at Upwood, which disbanded on 11 September 1961, although the Canberra was to continue in RAF service for many years in a variety of roles. The Canberra squadrons tended to be formed in Wings at their major bases, the bomber versions being the B.2 and B.6, while other Canberra squadrons served in Germany.

Binbrook, Lincs;
TA190960, nine miles SW of Grimsby by B1203

The airfield was built as a bomber station with five of the austerity 'C' Type hangars and technical site including workshops, stores, barracks, messes and administration buildings located in a group all on the edge of what was initially a grass landing ground. The incomplete Binbrook opened in June 1940 as part of No.1 Group Bomber Command with the Battles of the decimated 12 and 142 Squadrons arriving in July, having returned from France. Both squadrons moved out in August to allow building work to be completed. The squadrons returned to be equipped with Wellingtons, which went into operation in April 1941. In September 1942 the airfield was again vacated to allow the construction of three runways and connecting perimeter tracks to allow the operation of four-engine heavy bombers. The station reopened on 14 May 1943 with the arrival of 460 RAAF Squadron equipped with Lancasters, the squadron staying until the end of the war. The station commander was Group Captain Hughey Edwards, VC. The final mission of the war was on 25 April 1945 when 20 Lancasters set out to bomb Berchtesgaden.

In July 1946 9 Squadron arrived with Lincoln B.2s, and in September 617 (Dambusters) Squadron joined them, also with Lincolns, followed by 12 Squadron who moved in from Waddington. The four-Lincoln squadron Binbrook Wing was completed with the arrival of 101 Squadron in August, the crews working up on the aircraft before making a number of international goodwill trips, including 617 Squadron to the USA in 1947, these flights being good training in long-range navigation. From August to December 1949 the Lincolns of 50 Squadron were detached to Binbrook and 12 Squadron moved to Hemswell from January to March 1948, then returned to Binbrook. The withdrawal of the Lincolns began with 101 Squadron in June 1951 and the arrival of the first Canberra B.2. The remaining three squadrons began to replace their Lincolns with Canberras from early 1952, with the last Lincoln departed by May. The Canberra brought a whole new concept to RAF bomber operations and as more Canberras were delivered to Binbrook, 101 Squadron, led by Sqdn Ldr Cassidy, began conversion training of the aircrew, ready for intensive flying trials on 10 September, and by the end of November the squadron had ten aircraft. In addition to the flight trials, 101 Squadron included the Jet Conversion Flight and started training the crews for 617 Squadron, their first aircraft being delivered at the end of December 1951. While

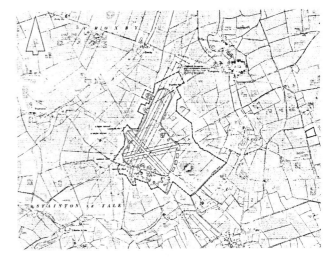

Below: The 1948 chart of Binbrook showed the standard three-runway bomber layout with an extension of the main runway to the north. The technical site included austerity 'C' Type hangars on the eastern side of the airfield and Lincoln B.2s arrived in July 1946, to be replaced by Canberras from June 1951. The bombers departed in December 1959, after which the station was transferred to Fighter Command with preparations for Lightnings. *RAF Museum*

617 Squadron continued to receive the new jet bombers, 101 Squadron continued the task of evaluating the Canberra under operational conditions, including intensive bombing and the evaluation of new equipment such as tail-mounted warning radar known as Orange Putter. The crews also flew long-range high-level navigation exercises up to 45,000 feet lasting some three and a half hours, and ending in a GCA approach to Binbrook. The first-generation fighters of Fighter Command often attempted to intercept these high-flying Canberras, but usually without success, often turning the tables on the Meteors by outmanoeuvring them. The Canberra could also outclimb the Sabre, and it was not until the entry into service of the Hunter that Fighter Command could intercept the Canberras. In April 1952 12 Squadron began to receive its Canberras, and the fourth was 9 Squadron with the first delivery in May; the fifth unit to complete the Binbrook Wing was 50 Squadron which began conversion in August. Some deliveries were delayed to allow 101 Squadron to relinquish the training task to 231 OCU at Bassingbourn, which remained the Canberra training unit for 15 years until the last Canberras departed for Cottesmore on 19 May 1969.

Meanwhile 101 Squadron continued to develop the operational capabilities of the Canberra, including the night intruder role and participating in Exercise Holdfast from Fassberg in Germany, followed by Exercise Ardent in which all RAF Commands had the designated task of probing the air defences of Britain, the Canberras being entirely unopposed in reaching their targets. Despite a second Canberra Strike Wing being formed at Scampton, in the Spring of 1953, the Binbrook Wing was still the only effective Main Force Canberra Wing,

Above: No. 617 Squadron was based at Binbrook from 1 May 1946 until 15 December 1955 when it was disbanded. The squadron operated Lincoln B.2s from September 1946 until January 1952, when it converted to Canberras. The squadron took its Lincolns on a goodwill tour of the USA in August 1947, including KC-C, KC-L and KC-B. *PHT Green collection*

taking part in Exercise Jungle King, the purpose of which was to test the defences of the Allied Tactical Air Forces in Germany and to provide a full-scale round-the-clock test of Bomber Command's readiness. The Canberras combined with Bomber Command Lincolns and Washingtons from 16 March initially at night, but also by day, many at low level. The Canberras were highly successful, while the Lincoln and Washington bombers were found to be very vulnerable, and had little chance of survival in a real war scenario. In May 1953, the Binbrook Wing took part in Exercise Rat/Terrier, where the Canberras were tasked with simulated low-level attacks against the USAF bases at Lakenheath and Sculthorpe to test their air defences, the defenders of Lakenheath being taken completely by surprise, and requesting a second attempt. Exercise also included attacks against seaborne targets simulating projected Soviet Naval threats. The ultimate bomber Canberra B.6 became available in June 1954 when it began to replace the earlier B.2s with 101 Squadron, with 9, 12, and 617 Squadrons receiving the improved B.6s during 1955. By the end of 1955 plans were in hand for the introduction of the V-Force with the reduction of the Canberra force, the first to go being 617

Left: No. 9 Squadron converted from Lincoln B.2s to Canberra B.2s in May 1952, which the squadron operated until June 1956. No 9 Squadron Canberra B.2s WD998 and WF907 were at Binbrook in 1953 during the changeover to the new jet bombers. *PHT Green collection*

Squadron which disbanded on 15 December 1955 later to reform at Scampton with Vulcan B.1s. No. 50 Squadron had already moved from Binbrook to Upwood on 8 January 1956 to be replaced by 109 Squadron which was disbanded on 1 February 1957. On that date 101 Squadron disbanded, some of its B.6s replacing the B.2s with 9 Squadron, which moved to Coningsby on 2 June 1959, together with 12 Squadron and 139 Squadron, which had moved from Hemswell to Binbrook in January 1956, disbanded on 31 December 1959. Binbrook was then reduced to C&M, but in April was transferred to Fighter Command with the main runway extended for the ultimate operation of Lightnings.

Coningsby, Lincs;
TF225565 South of Coningsby village by A153.

Well known as the home of the Battle of Britain Memorial Flight Lancaster, Hurricanes and Spitfires, Coningsby is now the major operational base for the RAF Typhoons. Coningsby was first opened on 4 November 1940 as a 5 Group Bomber Command airfield with the technical site including two 'J' Type hangars on the northern side of the grass landing ground, but after a short period of bombing operations, the station closed in the summer of 1942 for the construction of three concrete runways. It reopened again in August 1943 with an additional 'B1' hangar and three T.2 hangars, and the first resident was 617 Squadron with Lancasters, starting operations in mid-September with an attack on the Dortmund-Ems Canal led by Wing Commander Guy Gibson as the first of many bomber operations by a number of squadrons, the final one being on 25/26 April 1945 by 83 and

Above: The 1951 plan of Coningsby shows the three-runway layout with the technical site on the north side of the airfield including two 'J' Type hangars with three T.2 hangars behind and a fourth by the external road, which is now the home of the Battle of Britain Memorial Flight. There was also a large weapons storage area to the south of the airfield which is now the site for a batch of HAS. Lincolns were based at Coningsby from July 1948, although the station closed in March 1950 to make it ready for Canberras, but in the meanwhile it was one of the two RAF Washington bases. In 1953 the Canberras began to arrive, but the airfield was closed again in early 1954 for the main runway to be extended to 9,000 feet, reopening at the end of 1956. From July 1961 preparations were made for the V-Force and Coningsby is now the home for Typhoons. *RAF Museum*

No 57 Squadron operated Canberra B.2s from May 1953 until December 1957, and was based at Coningsby from November 1956 until disbanding in December 1957. One of their Canberras was WK131 which was retired to Halton for ground instruction as 7912M, where it was seen in September 1970. *Author*

Above: The 149 Squadron crew parade in front of their Washington B.1 at Coningsby on 3 July 1951. *Andrew Thomas collection*

Cottesmore, Leicester;
SK905155 W of A1 and N of B668 by Cottesmore village.

In what was then the County of Rutland, construction of Cottesmore started in 1935 as part of the RAF Expansion Scheme as a bomber station within 2 Group to accommodate two squadrons. Although it was not fully ready, the station opened on 11 March 1938 with the technical site on the south side of the landing ground, consisting of four 'C' Type hangars in a gentle semicircle with the stores, workshops, administration, messes and living accommodation between the airfield and village. The first flying units were 35 Squadron with Wellesleys and 207 Squadron with Battles, both types being somewhat outdated. The station then became responsible for the training of bomber crews, some of the more experienced crews being used on operational raids, including participating in the 1,000-bomber raids in mid 1941. In August 1943 the station was placed on C&M in preparation for the construction of three concrete runways, the main one being 2,000 yards long joined by a perimeter track with 35 concrete and 17 tarmac dispersals. In addition to the original 'C' Type hangars, a single T.2 was erected at the SW end. The station was taken over by the USAAF on 24 September 1943 as the HQ of Troop Carrier Command, and although there were initially no regular flying units, the station was used as a transit base for American ground units. On 16 February 1944 the first C-47 Skytrains, C-53 Skytroopers and Hadrian gliders arrived as part of the build-up of the planned invasion force, the preparation and training continuing until 72 aircraft from Cottesmore were tasked with the dropping of paratroops and the launching of gliders on Operation Overlord, the Allied invasion of Europe on 6 June 1944. Following the initial landings, the units at Cottesmore continued with troop and supply flights in support of the advancing armies, and in September were also involved in Operation Market Garden, the ill-fated airborne landings at Arnhem.

With the end of the war in Europe the Americans had all departed by 11 June 1945 and the station was handed back to the RAF. Cottesmore was allocated to the training role, which continued until May 1954 when the station was transferred to 3 Group Bomber Command. The Coningsby Wing of 15, 44, 57 and 149 Squadrons arrived with Canberra B.2s in May 1954, but plans were in hand to convert Cottesmore to V-Bomber operations, and on 24 August 1954, 149 Squadron moved to Ahlhorn in Germany as part of the 2nd TAF, and the other three squadrons moved out to Honington in early February 1955, the station being put under C&M on 15 February ready for the major changes for the operation of Victors.

97 Squadrons. Both squadrons remained at Coningsby after World War 2 and re-equipped with Lincoln B.2s in July and August 1946, and in November they moved out to Hemswell. The station then became home for Mosquitos of 109 and 139 Squadrons, in addition to 16 OTU and 231 OCU, until March 1950 when both the squadrons moved to Hemswell in preparation for conversion to Canberras.

With the Lincoln becoming obsolete, Coningsby became one of the two main bases for the US-supplied Washington long range heavy bombers, the first to arrive being for 149 Squadron in October 1950, having equipped and worked up at Marham. The Washington heavy bomb wing built up with the arrival of 15 Squadron in February 1951, 44 Squadron in May and finally 57 Squadron in April 1952. However, the Washington was also becoming ineffective and in 1953 the four resident squadrons converted to Canberras, starting with 149 Squadron in March, and the Wing was expanded with the arrival of 40 Squadron in October. In February 1954 40 Squadron moved to Wittering and the remaining four resident units moved to Cottesmore to allow for major building improvements including the extension of the main runway to the east with an ORP at the end, the closing of the N/S runway and the construction of a Gaydon Type hangar. When Coningsby reopened in late 1956 the new runway was 9,000 feet long and 200 feet wide making the airfield suitable for the jet age. Canberras returned, the first being 57 Squadron from Honington in November 1956, disbanding on 9 December 1957. On 2 June 1959 9 Squadron arrived and were joined by 12 Squadron on 2 July. On 13 July 1961 both squadrons disbanded in preparation for re-equipping with Vulcan B.2s as part of the V-Force.

Hemswell, Lincs;
SK940910, approximately 13 miles N of Lincoln, W of A15

The site close to the Roman Road of Ermine Street was opened as a training airfield in June 1918 named after the village of Harpswell on the southern boundary, but it was closed after a year and returned to agriculture. However, in the early 1930s, the site was surveyed and it was selected as one of the early Expansion Scheme bomber bases. The technical site was constructed on the eastern boundary of the grass landing ground with a semi circle of four 'C' Type hangars, behind which were the workshops, administration buildings, messes and living accommodation. The new airfield, now known as Hemswell after the village on the western boundary, opened in January 1937 as part of 5 Group Bomber Command. The first flying units were 144 Squadron in February 1937, soon joined by 61 Squadron, their equipment being Hampdens when war was declared. No. 61 Squadron were the first unit in Bomber Command to drop bombs on German soil on 19 March 1940 when the target was Hornum seaplane base. Bombing operations continued until June 1943 when the airfield closed to allow the construction of three concrete runways with perimeter track giving access to dispersals. The work was completed by January 1944, and in October Hemswell became a satellite to Scampton with the Lancasters of 150 Squadron arriving on 11 November, followed by 170 Squadron on 30 November. Wing Commander Guy Gibson, leader of the Dams Raid with 617 Squadron, was commander of 150 Squadron on Pathfinder duties and he was flying from Hemswell when he was killed with his navigator in a Mosquito on 19 September 1944 while returning from Master Bomber duties during a raid on Rheydt. Bombing operations continued until April 1945, both squadrons disbanding in November. Although many airfields were closed as surplus to requirements, Hemswell was retained and during October and November 1946 83, 97 and 100 Squadrons arrived with Lincoln B.2s. Various other Lincoln units were detached from time to time. In March 1950 100 Squadron moved out to Waddington, and both 83 and 97 Squadrons were the last to operate the Lincoln, when they both officially disbanded on 1 January 1956. The bombing role continued with the arrival of 109 and 139 Squadrons in March 1950 initially equipped with Mosquito B.35s, and then replaced by Canberra B.2s between August and November 1952. The squadrons then received Canberra B.6s and both units moved to Binbrook on 1 January 1956. The Canberra B.6s of 76 Squadron were resident from 1 April 1957 until they departed to Upwood on 17 July 1958. The airfield then closed to flying and was used from December 1959 until May 1963 as a Thor ICBM base with 97 Squadron, Fighter Command. A total of nine US-supplied Thor missiles were mounted on mobile launchers and each was tipped with a one-megaton nuclear warhead, under the joint control of the RAF and USAF. The Cuban Missile Crisis brought the entire UK-based Thor missile inventory to maximum readiness for a ten-day period during October and November 1962. The Station finally closed in early 1967, and although the runways have been removed, much of the original technical site still exists in commercial use.

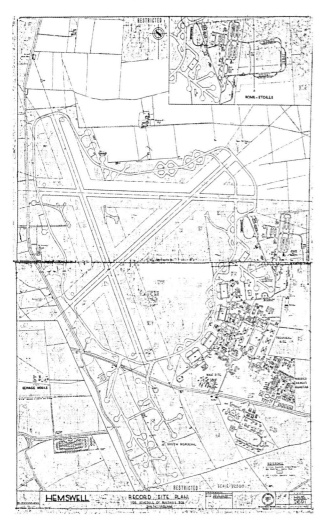

Left: Hemswell was one of the earliest of the RAF Expansion Scheme stations, and opened in January 1937 as an RAF bomber station with the technical site on the southeast side of the airfield. The traditional semi-circle of four 'C' Type hangars with a watch office in the centre faced the airfield with workshops, stores, administration and accommodation spread out behind the hangars. This included 'H' block barracks, WAAF accommodation and Airmen's married quarters. In June 1943 the airfield was closed for the construction of three runways with perimeter track and aircraft dispersals. An additional T.2 hangar was erected to the north of the existing hangars with an extensive weapons store to the east. Combat operation continued until April 1945 and in October 1946 Lincoln B.2s arrived staying until January 1956. Canberras arrived in March 1950, and the airfield ceased operations in August 1958. The station closed in early 1967 and although the runways and associated concrete has been removed, much of the technical site remains in commercial use. *RAF Museum*

Left: No 97 Squadron Lincoln B.2s, including RF526, ready for departure at Hemswell in April 1948. The squadron operated Lincoln B.2s from July 1946 until December 1955, and it was based at Hemswell from 7 November 1946 until it was disbanded on 1 January 1956. *IWM HU65271*

Honington, Suffolk;
TL890755, 7 miles NE of Bury St Edmunds between A134 & A1088.

Honington was one of the earliest permanent RAF stations built under the Expansion Scheme, announced in July 1934, and was built to a very high standard. Site preparation commenced in 1934, ready for construction to begin in 1935, and opened in July 1937 as a bomber station in 3 Group. The technical site was on the south side of the landing ground with four 'C' Type hangars in a semicircle with a fifth behind the westerly one. The accommodation included 'H' block barracks, station HQ, stores and workshops behind the hangars. The first flying units were 77 and 102 Squadrons with Audaxes and at the outbreak of war the main types were 9 Squadron with the first Wellington bombers from mid July 1939 until departure on 7 August 1942. They were joined by 215 Squadron on 8 April 1940 and 311 (Czech) Squadron formed at Honington on 29 July, the Wellington operations continuing until 22 August 1942, ready for the arrival of the Americans. The station became a major modification and overhaul centre for B-17s of the 3rd Air Division and on 29 April 1943, Honington officially passed from RAF to USAAF control. It was designated the 1st Strategic Air Depot on 6 September 1943, able to accept B-17s damaged on operations if their undercarriage would not lower and a crash landing was unavoidable, which was easier than bringing the damaged aircraft in by road. By 29 June 1944, some 400 B-17s had crash-landed at Honington and either been repaired or stripped for spares recovery.

The landing area was still grass surface, and to accommodate the 364th Fighter Group P-38 Lightnings which arrived on 10 February 1944 for defence escort of the USAAF day bombers, a 6,000ft steel mat runway was laid. In addition taxiways were constructed with 75 hardstandings, and nine blister hangars were erected together with other temporary buildings. The P-38s also undertook the hazardous strafing of German airfields and provided top cover for the Allied ships preparing for the D-Day landings. The more potent P-51D Mustang began to replace the Lightnings, the last P-38 operation being on 29 July 1944. The P-51Ds continued in operation covering the Battle of the Bulge and Rhine Crossing, the final mission being on 25 April 1945. Honington was retained by the American forces until 26 February 1946 when the RAF regained control and it became part of Transport Command

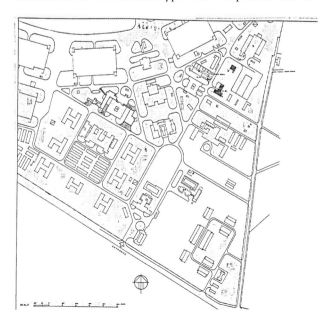

Left: The technical site at Honington has changed very little since 1955 when the plan was drawn. The station was one of the early RAF Expansion Scheme stations, work beginning in 1935 ready for opening in July 1937 as a 3 Group Bomber Command base. The guard room was located on the right of the main entrance with the station HQ facing the gate. 'H' barrack blocks provide airmen's accommodation and the five 'C' Type hangars with the watch office at the centre face the airfield. A steel-mat runway was laid in 1944, but it was not until September 1954 before a 9,000ft concrete runway was constructed ready for Canberra jet bomber operations from February 1955. The Canberras departed in July 1957 and the station was made ready for the V-Force in the strategic bomber role. *RAF Museum*

Left: Based at Honington, 15 Squadron operated Canberra B.2s from May 1953 until April 1957. The squadron was latterly based at Honington from February 1955 until it was disbanded on 15 April 1957. Canberra B.2 WK132 with the Honington Canberra Wing speedbird on the fin was retired to Halton as 7913M for ground instruction, where it was seen in September 1970. *Author*

as the Major Servicing Unit, supporting RAF Dakotas on the Berlin Airlift.

The station returned to Bomber Command in 1949 and in 1950 was used by 94 MU for the storage of ammunition and bombs. Barnham Camp, to the northwest of Honington near the A11, was established in June 1954 as a special storage area for atomic weapons, and in September 1954, 94 MU moved there allowing Honington to be enlarged and a concrete runway laid for the first time, 6,000 feet long and 200 feet wide for jet operations. The first Canberras of 15, 44 and 57 Squadrons arrived in February 1955, the Wing being completed with the arrival of 10 Squadron from Scampton in May. Some of the Canberras were detached to Cyprus for Operation Musketeer in the late summer of 1956, 10 and 15 Squadron each sending eight Canberras. Two crews from each squadron participated in the initial bombing raid on Cairo Airport on 26 October. A total of 28 sorties were flown by the Canberras during the Suez Canal emergency, but soon after their return to Honington with the build-up of the V-Force, the Canberra force began to wind down.

On 15 November 1956, 57 Squadron moved out to Coningsby, 10 Squadron disbanded on 1 January 1957 and 15 Squadron disbanded on 14 April followed by 44 Squadron on 15 July, all to re-equip with V-Bombers. Honington then prepared for the arrival of the new RAF strategic bomber force with extensions to the main runway in a westerly direction and an ORP at the eastern end.

Marham, Norfolk;
TL730685 10 miles E of Downham Market,
N of the Swaffham Road.

Marham opened in 1916 on an 80-acre site on the eastern part of the present airfield, its role being night home defence by 51 Squadron, but the site closed in May 1919, returning to agriculture. The site was requisitioned as part of the RAF Expansion Scheme, which provided permanent stations for the planned bomber force, while the fighter stations were relegated in many cases to World War 1-vintage facilities. The new Marham opened on 1 April 1937 as a two-squadron heavy bomber base with four 'C' Type hangars in a semicircle on the north side of the landing ground, and a fifth similar hangar behind the most northerly one. The extensive technical site behind the hangars consisted of the station HQ, stores, workshops and living accommodation on a slight hill overlooking the airfield. The first flying unit was 38 Squadron with Fairey Hendon bombers, the first of which arrived on 5 May, but the squadron began to convert to Wellington bombers in November 1938. The second unit was 115 Squadron, which formed on 15 June 1937, initially with Harrows, which began to be replaced by Wellingtons in April 1939. The first action from this bomber base was 3 December 1939 when Heligoland was the target, and on 28 August 1940, the Wellingtons made their first night attack on Berlin, bombing operations continuing with the type until March 1942, when the resident 218 Squadron re-equipped with Stirlings.

In August 1942 Marham became part of 2 Group Bomber Command and 105 Squadron, responsible for introducing the

Below: Marham closed in April 1944 for the construction of three concrete runways, perimeter track and bomber hardstandings and did not reopen until mid-1945, when it was prepared as one of the two RAF Washington bases. The technical site with five 'C' Type hangars is on raised ground in the northwest corner of the airfield, and this 1944 vertical shows the new runways under construction. *IWM C5469*

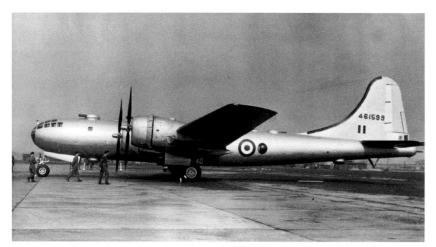

Left: The first ex USAF B-29 461599 for the RAF as a Washington was handed over at Marham in February 1950, and although it carried RAF insignia, the British serial was yet to be allocated. *RAF Museum P.013497*

Mosquito high-speed unarmed wooden bomber to the RAF arrived on 29 September, to be joined by 139 Squadron on the same day. These two squadrons specialised in low-level daylight precision attacks. One of the most memorable attacks was to celebrate Hitler's birthday on 30 January 1943, when 105 and 139 Squadrons made two attacks on Berlin, 105 Squadron dropping the first bombs at 11.00 hrs delaying a propaganda speech by an hour and returning safely without being intercepted. Three aircraft of 139 Squadron bombed at 16.00 hrs, but the defending fighters were ready and shot down the leader, Sqdn Ldr Darling. The two Mosquito squadrons were transferred to 8 Group in June 1943, making nuisance raids over Germany at night to keep the defences on alert. With the development of Obeo, the squadrons specialised in precision pathfinding for the main night bomber force. In July 1943 139 Squadron was replaced by 109 Squadron, and both squadrons continued on pathfinding duties until April 1944 when the station closed to allow a 9,000ft long by 300ft wide runway to be constructed for the operation of very heavy bombers, putting Marham out of operation until the end of the war.

Marham reopened in the summer of 1945 with two subsidiary concrete runways, interconnected by a concrete perimeter track and 36 heavy bomber dispersals. On 25 September the Central Bomber Establishment moved in conducting trials until February 1950 when plans were made for the RAF introduction of ex-USAAF B-29 interim heavy bombers, the first example arriving in April. With the Lincoln incapable of delivering an atomic bomb, there was a need for something more capable in the interim before the arrival of the Canberras and later the V-bombers. The USAF was in the process of replacing their B-29s with the improved B-50s and the larger B-36, making some B-29s available; their high-altitude pressure cabins were a great improvement over the Lincolns. Therefore under the US Military Aid Programme (MAP) the RAF acquired 87 surplus B-29s for RAF Bomber Command, designated Washington B.1s, which were enough aircraft to equip nine squadrons with eight aircraft each.

On 1 July the Washington Conversion Unit (WCU) formed for the training of the crews, starting with 35 Squadron, and followed by 15, 44, 90, 114, 149 and 207 Squadrons. The Washingtons of 115 Squadron, which was the first to convert in the USA, brought their first aircraft to Marham at the beginning of March 1950 plus those of 35, 90, and 207 Squadrons which remained at Marham, while the remainder moved to Coningsby. Marham became very busy when they were joined by USAF B-29 squadrons during the Korean War, the first arriving in August 1948 from the 28th and 307th BGs, and the last departing in February 1951. Included in these deployments were USAF B-50s with a capability of carrying an atomic bomb, and the American bombers came under the control of the newly-formed 3rd Air Division USAFE with HQ at Marham.

As the lead RAF unit, 115 Squadron participated in many air defence exercises and in 1951 and 1952 the operations included the development of operational radar bombing techniques, while in 1953 115 Squadron was awarded the Laurence Minot Trophy for its bombing expertise. Bomber Command Washingtons were mainly tasked with heavy bombing training operations in preparation for the start of WW3, centred on multi-engine pilot skills, bomb-aiming for conventional and nuclear actions, navigational skills, gunnery perfection and flight engineer training. Equally, the ground crews gained their skills on the maintenance of more modern airframes and engines. Marham-based Washingtons participated in Exercise Emperor in October 1950 with the USAF B-50s of the 93rd Bomb Group. In July 1952 90 Squadron Washingtons won the RAF Bomber Command bombing competition, the main part of the competition being four cross-country night flights and visual bombing on a range. Points were awarded for bombing accuracy and navigation skills, the achievement leading to the award of the Laurence Minot Trophy again for a Marham squadron. The RAF Washingtons were only used as an interim long-range heavy bomber until sufficient Canberras were available, the last Washington being retired from 207 Squadron in March 1954.

Meanwhile in February 115 Squadron rearmed with Canberra B.2s, followed by 207 Squadron in March, 35 Squadron in April and 90 in November to form the Marham Canberra Wing. With the arrival of 214 Squadron Valiants in March 1956, 207 Squadron disbanded on 27 March to fly Valiants; 90 Squadron disbanded on 1 May and 35 Squadron departed for Upwood with its Canberra B.2s, making Marham a fully operational Valiant station.

Mildenhall, Suffolk;
TL685768, 12 miles NW of Bury St Edmunds, by A1101.

Just after opening on 16 October 1934, Mildenhall's early claim to fame was as the starting point for the MacRobertson Air Race to Melbourne in Australia, the race starting on 20 October, and the winner being the specially designed and built de Havilland Comet Racer sponsored by the London Hotel, Grosvenor House. The winning Comet G-ACSS still exists in flying condition with the Shuttleworth Trust at Old Warden.

The go-ahead for the building of an RAF bomber base was not given until 1933, when the Expansion Scheme had hardly become established, so the construction was not to the traditional layout. The technical site was constructed on the north side of the site and initially two Type 'A' hangars were built, later followed by three 'C' Type. The Station HQ, main guardroom and officers' mess were similar to traditional Army-style brick construction. After the air race to Australia had departed, the first RAF aircraft to arrive were the Heyfords of 99 Squadron, later splitting to add 38 Squadron and in April 1937 forming 149 Squadron also with Heyfords. By the time war was declared, the squadrons were equipped with Wellingtons with 149 Squadron and 99 Squadron dispersed to Newmarket. The squadrons participated in the night bombing offensive to the end of 1942, Stirling heavy bombers being introduced in October 1941. The station closed in November 1942 for the construction of three runways in the standard pattern with 36 heavy bomber dispersals around the perimeter track, ready for 15 Squadron to return with Stirlings on 15 April 1943, later to form 622 Squadron. Both squadrons re-equipped with Lancasters in December 1943, continuing on night operations until the end of the war. After the war Mildenhall continued in the bombing role with the Lancasters of 35, 115, 149 and 207 Squadrons arriving in February 1949 ready for conversion to the Lincoln. In July 207 Squadron was the first to receive Lincoln B.2s, all four units having re-equipped by October. However, the aircraft were already outdated, and on 23 February 1950 35 Squadron disbanded, followed by the other three on 1 March. The airfield was then upgraded with the main EW runway extended in both directions in early 1951, having reopened on 15 July 1950 to accommodate the B-50s of the 93rd Bomb Group USAF, the base being under American control since that time.

Scampton, Lincs;
SK965795 6 miles N of Lincoln by A15.

Scampton's main claim to fame is as the base of 617 Dambuster Squadron which lead by Wing Commander Guy Gibson breached the Mohne and Eder Dams on the night of 16/17 May 1943 using specially modified low-level Lancasters. They placed the special weapons against the inside of the dam wall, with a hydrostatic trigger set to explode the mine against the structure. Of the 19 Lancasters that set out, nine failed to return, with the loss of 53 men.

The site was first used in World War 1, at that time known as Brattleby, and opened in November 1916 as a training station with mainly temporary buildings. It was also used for local air defence against Zeppelins at night, without any aids, and no chance of intercepting an enemy airship as they flew over at 18,000 feet, while the defending fighters could barely reach 12,000 feet. Flying operations ceased in April 1919 and the site was returned to agriculture by 1920 with all trace of the airfield gone.

In 1935, as part of the RAF Expansion Scheme of constructing bomber airfields, work started on an enlarged World War 1 site with a grass landing ground and the technical site on the SE corner of the camp. There was the usual semicircle of four 'C' Type hangars with the workshops, stores, administration, messes and accommodation between the hangars and the A15 road. The new station opened as RAF Scampton on 27 August 1936 within 3 Group Bomber Command, although there was

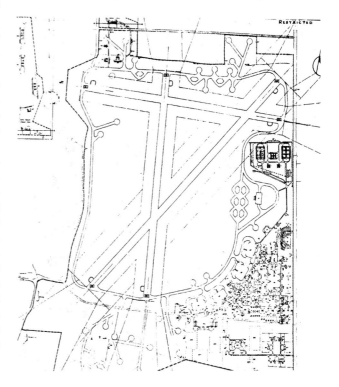

Above: In 1946 Scampton retained its World War 2 layout which had originally been opened in August 1936 with four 'C' Type hangars as part of 3 Group Bomber Command. The station was closed in August 1943 for the construction of the three runways with perimeter track, dispersals and other improvements, ready for operations to restart in September 1944. After World War 2 Canberras were based at Scampton from January 1953 until May 1955. The station was then put under C&M while preparations were made for V-Bomber operations; the main NE-SW runway was extended to 9,000 feet to the northeast, resulting in a diversion of the A15 Roman Road in a wide loop. *RAF Museum*

Above: No 57 Squadron was resident at Scampton from 2 December 1945 until 1 May 1946 and flew Lincoln B.2s from November 1945 to May 1951. Lincoln B.2 DX-L:57 Squadron with at least one more aircraft was obviously deployed to somewhere warmer than Scampton. *RAF Museum P.015405*

still a considerable amount of construction to be completed. However, in October the first flying units arrived, 9 Squadron with Heyfords and 214 Squadron with Virginia and Harrows before departing in April 1937. The bomber units at Scampton built up ready for conflict, each squadron having 16 aircraft with five in reserve, and the construction was completed in September 1939, in time for the declaration of World War 2 on 3 September. Operations were started gradually by 49 and 83 Squadrons initially flying Hampdens until the spring of 1942. In July 1941 both squadrons had there strength increased to 26 Aircraft each split into three flights, and in December 1941 the first of the unsuccessful Manchester bombers arrived for 83 Squadron. The Vulture engines powering the Manchester were very unreliable and by May the Manchesters were being replaced by the Merlin-powered Lancasters. Meanwhile 49 Squadron operated the Manchesters until they were also replaced by Lancasters. The two Scampton squadrons took part in the three 1,000-bomber raids, the first being on the night of 30/31 May 1942 against Cologne, and after the third raid against Bremen on 25/26 June, the Manchester was finally withdrawn from operations. Bomber operations continued, including the famous Dams raid, before the station closed in late August 1943 for a nine-month programme of improvements, including the construction of three concrete runways. By the end of September 1944 the construction work was complete and the following month Lancasters of 153 Squadron arrived, the base being transferred to 1 Group with bomber operations continuing until the end of the war and 153 Squadron disbanding on 28 September 1945. No. 57 Squadron operated Lincoln B.2s at Scampton from 2 December 1945 until 1 May 1946.

Following a period in the bomber training role, Scampton was allocated to the US Government and on 17 July 1948 30 B-29s of the 28th Bomb Group arrived as part of the defence against the increasing tension from the Soviets. As the crisis reduced, on 15 January 1949 the B-29s returned to the USA, and preparations were made for the arrival of Canberra jet bombers, bringing the station back into the bomber role. The first to reform was 10 Squadron with Canberra B.2s on 15 January 1953, and the Wing was completed by the reformation of 27 Squadron

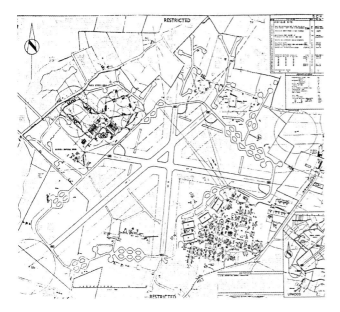

Left: Upwood opened in January 1937 with four 'C' Type hangars fronting the technical site on the southeast boundary of the airfield. On the left inside the main entrance was the guardroom with the station HQ opposite. Airmen's accommodation included brick built 'H' barracks and the station entertainment included a cinema. The station was closed from April 1943 to January 1944 for the construction of three concrete runways, perimeter track and 36 bomber hardstandings. After World War 2 the station was allocated to Transport Command, but returned to Bomber Command in July 1946. The Lancasters were replaced by Lincoln B.2s in September 1949 which were then withdrawn by June 1952. Canberras arrived in May 1955 and remained with the Upwood Canberra Wing until January 1959. All flying ceased at Upwood in September 1961, following which the station closed apart from a USAFE clinic being built on the eastern part of the technical site in support of Alconbury. Although the runways and associated concrete have been removed, the technical site remains largely intact. *RAF Museum*

on 15 June, 18 Squadron on 1 August and 21 Squadron on 21 September. The Canberras remained until May 1955, by which time Scampton had been designated a V-Bomber base, and on 1 June the station was closed and put under C&M in preparation for Vulcan bombers, including the extension of the main runway to 9,000 feet at the NE end, involving the diversion of the A15 from its straight Roman route.

Below: Canberra B.2 WH723 was located on the gate of Upwood, painted as WJ642 with 35 Squadron markings. Seen in August 1967, it has since been scrapped. *Author*

Upwood, Cambs;
TL270845 2 miles SW of Ramsey by B1096 road

Aviation first came to the site as a night flying training camp from July 1918 until 26 June 1919. The site was selected as an RAF bomber station in the mid 1930s under the Expansion Scheme with the technical site on the SE corner, including three 'C' Type hangars, with a fourth located behind the westerly one. The remainder of the station buildings, including the Station HQ, Officers' and Sergeants' messes and guardroom were all built from bricks. Following opening in January 1937, it became the first base for the Fairey Battle bomber when 63 Squadron received the first of what proved to be a rather vulnerable and underpowered aircraft in May 1937. The Hinds of 52 Squadron were also replaced by Battles at the end of the year. With insufficient range to bomb Germany from Britain, Battles were deployed to France when war was declared as part of the Allied defences, where they aircraft were effectively destroyed with the loss of many crews. Upwood then became an operational training base with Blenheims until April 1943, when flying ceased to allow construction of the normal pattern of three runways connected by a perimeter track with 36 heavy bomber loop hardstandings.

The station opened again on 31 December 1943 and at the beginning of February 1944 the Mosquito B.XXs of 139 Squadron arrived equipped with H2S radar to carry out target marking for other Mosquito squadrons. In March 1944 the Lancasters of 156 Squadron joined the Mosquitos, providing target marking and bombers for the Pathfinder Force. On 27 June 1945, 156 Squadron moved to Wyton and 105 Squadron took their place, making Upwood an all-Mosquito

Above: No 50 Squadron operated Canberra B.2s at Upwood from 8 January 1956 until it was disbanded on 1 October 1959. Canberra B.2 WJ635 with 50 Squadron was at Upwood in 1957. *Michael Bowyer*

base. With the departure of the bombers, Upwood then came under the control of Transport Command from February 1946 with 53 Squadron flying Liberator transports for troops returning from India. The squadron converted to Yorks in April, and was disbanded on 25 June. The station was then allocated to Bomber Command and was refurbished for the operation of heavy bombers, including modest extension to the runways ready for the Lancasters of 49 Squadron which arrived on 29 July 1946. They were joined by 7 Squadron the next day and on 4 November by 148 and 214 Squadrons returning from Egypt. Starting in 1949 the four squadrons converted to Lincoln B.2s, beginning with 7 Squadron in September and finishing with 214 Squadron in February 1950. The Lincolns were used on overseas deployments in Malaya, Aden and Kenya, 214 Squadron disbanding on 30 December 1954, 148 on 1 July 1955 and 7 Squadron on 1 May 1956, 49 Squadron having moved to Waddington on 25 June 1952.

In place of the Lincolns came the Canberra B.2s starting with 18 Squadron which arrived on 22 May 1955 as the first of four squadrons to form the Upwood Wing. They were followed by 61 Squadron on 30 June 1955, 50 Squadron on 8 January 1956 from Binbrook and 35 Squadron on 16 July. In late October and early November 1956, Canberras of both 18 and 61 Squadrons took part in the Suez Campaign as part of the Cyprus Bomber Wing based at Nicosia, making four night bomber attacks. With the build-up of the V-Force, the Canberra units began to reduce with 40 Squadron, which had arrived on 31 October 1956, amalgamated with 50 Squadron on 15 December. On 1 February 1957 18 Squadron disbanded, followed by 61 Squadron on 31 March 1958. In October

Left: Although the airfield has been returned to agriculture the technical site still exists, including four "C" Type hangars, the station HQ and guard room. *Author*

1958, 21 Squadron arrived from Waddington, but disbanded on 15 January 1959. The airfield was closed to flying on 11 September 1961, and part of the technical site was taken over by the USAF at Alconbury as a clinic. Most of the runways have been removed, and the airfield returned to agriculture, apart for a section of grass alongside the main runway which is used for gliding based in the SW corner. Much of the camp still exists in good condition, although one of the hangars was converted into a mortuary in case of high casualties from the First Gulf War. Another hangar has been used for commercial purposes, initially on aero engine overhaul, the site remaining as a reminder of what a 1950s RAF station was like.

Waddington, Lincs;
SK985645 4 miles S of Lincoln between A.15 and A.607

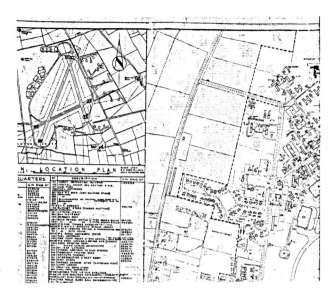

Above: Situated alongside the A15 with its Roman origins and a few miles south of Lincoln, Waddington reopened in October 1926 and was used for flying until the mid-1930s when it was developed into an RAF bomber station to the RAF Expansion Scheme standards. The technical site was located in the northwest corner with five "C" Type hangars, but not to the normal semi-circle layout, the two centre ones being aligned, and the fifth hangar behind the northerly one. The extensive technical site includes the usual technical, administration and accommodation buildings including "H" barrack blocks. A weapons storage site was located in the southeast corner. The station was closed in May 1943 for the construction of three concrete runways with perimeter track leading to dispersed hardstandings for bombers which was completed by November 1943. The final wartime operation was on the night of 25/26 April 1945, and the station was retained by Bomber Command, with Lincoln B.2s arriving from May 1946. The station closed in August 1953 for preparation as a V-Bomber base with the main NE-SW runway extended to 9,000 feet. Amongst other developments, a nuclear weapons store was built on the east side of the airfield over the A15. The station is still active, and developments include a new hangar for the Sentries to the north of the original technical site. The 1954 site plan shows Waddington as it was being developed to accommodate the V-Force. *RAF Museum*

The station first opened in November 1916 in the flying training role, which continued until 1919, when Waddington closed. However, it was one of the few airfields to retain its landing ground and buildings during the run-down after World War 1, and it reopened in October 1926 as a reserve base until 1 May 1936 when 503 (Bombing) Squadron became an Auxiliary unit equipped with Fairey Fawn light bomber. During the mid-1930s a rebuilding programme was commenced along the lines of the RAF Expansion Scheme with five 'C' Type hangars on the west side of the airfield and the technical site behind, the main entrance being off the A.607 road. On 3 May 1937 50 Squadron re-formed at Waddington with Hinds, followed by 110 Squadron on 18 May and 88 Squadron on 7 June. In June 44 Squadron arrived, soon to equip with Blenheims, and 88 Squadron departed. By the start of World War 2 both 44 and 50 Squadrons were flying Hampdens, with the first operation by nine aircraft of 44 Squadron on the day war was declared. The Hampdens were in front-line operations until 1 November 1940 when 207 Squadron re-formed with the twin-Vulture-engine Manchester bombers, which was plagued with development problems and was replaced at Waddington by Lancasters for 44 Squadron from 24 December 1941. Bombing operations continued until May 1943, when it closed for the construction of the standard three concrete runways linked by a perimeter track with bomber dispersals around the boundary. The station reopened in November 1943 when night bomber operations continued, 467 (RAAF) and 463 (RAAF) Squadrons being the new residents. The squadrons played a major part in support of the D-Day landings with attacks on the shore gun batteries, and their last operational mission of the war was on the night of 25/26 April 1945.

Waddington continued as a bomber station after the war, and in May 1946 61 Squadron re-equipped with Lincoln B.2s. In July 12 Squadron arrived from Binbrook to re-equip with Lincolns, before returning there in September. In October 57 Squadron arrived from Lindholm with Lincolns, having been undertaking trials with early production aircraft, and in March 1950 100 Squadron arrived staying until 2 August 1953 when it moved out to Wittering, together with 49 and 61 Squadrons. The Lincolns equipped the new No.1 Group Bomber Command with the HQ at Bawtry, all being based in Lincolnshire and each squadron had a strength of eight aircraft, half of the World War 2 strength. No.57 Squadron, together with the other early Lincoln-equipped units participated in the development of the aircraft, including routine high-level visual and radar bombing, air-to-surface gun firing, day and night cross-country navigation exercises, instrument approaches and systems operation. Other duties included weather reconnaissance flights when it was found that during long cold flights the plastic fuel pipes contracted, pulling the ends off the connections, in some cases causing the loss of two engines on one side and half the fuel. In comfort the Lincoln was little better than the Lancaster with long, cold, boring night exercises in the winter with no heating causing much discomfort. I had the dubious experience of flying as an air cadet in a Lincoln from Manby on my third flight in one day. It was a three hour, 20 minute flight over Northern Ireland and Scotland at 33,000 feet in the mid-upper astrodome on oxygen continually. As the aircraft was unpressurised, the long, slow approach to land was extremely painful in the ears, the final touchdown being at night.

Waddington did not accommodate a Canberra wing, as it was closed in August 1953 for preparation as a V-bomber base with the main runway extended to 9,000 feet, and the other two runways withdrawn from use.

Wittering, Cambs;
TF045025, 3 miles S of Stamford alongside the A.1

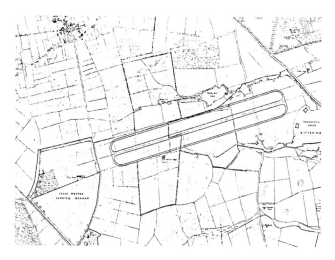

Above: Probably one of the oldest RAF stations, Wittering is located alongside the A1 south of Stamford, after which the airfield was named when it originally opened December 1916. The name Wittering, after the nearby village, was adopted in May 1924, and from 1935 the technical site at the eastern end of the airfield was developed to include three 'C' Type hangars. The station reopened in May 1938 as a fighter base, which continued throughout World War 2 and the first runway was constructed in 1941. In 1945 the airfield was combined with nearby Collyweston to the west, and the runway extended. The airfield closed in April 1950 for major developments, including the extension of the runway to 9,052 feet and a Gaydon hangar was erected on the technical site. Bomber Command took over the station in 1953 and Lincoln B.2s arrived in August with the first Blue Danube atomic bomb being delivered in November. Canberras arrived in December 1953, and remained until September 1959, when preparations commenced for the operation of V-Bombers. *RAF Museum*

Known as the home of the RAF Harriers, Wittering has had a long and varied history. The site was first used as a home defence station known as Stamford from December 1916 and became a storage depot at the end of 1919 until placed under C&M in January 1920. The airfield was resurrected and became home for the Central Flying School (CFS) from 17 October 1926, the name Wittering being adopted in May 1924. Because of its early history, the station layout is non-standard with the technical site in the SE corner of the flying field, and many of the early buildings remain in use, the most obvious example being the original guardroom just off the A1. The original hangar accommodation was a pair of Aircraft Shed Doubles one of which was used in half as a workshop, and the other half as a store. The site included Station HQ, main stores, new Sergeant's Mess and four barrack blocks. The Officer's Mess is 1924 style and the Airmen's Married Quarters are a 1922 design. Flying training continued until May 1938, by which time the site had been further developed with work starting in 1935. One of the original hangars was retained, three nine-bay 'C' Type hangars were constructed and the existing buildings were upgraded. Around the grass landing ground a concrete perimeter track was laid with ten dispersals. The station reopened on 16 May 1938 as a fighter base with the Demons of 23 Squadron and Gauntlets of 213 Squadron. During the Battle of Britain Wittering was a sector station in 12 Group, providing local air defence and support to the hard-pressed 11 Group squadrons in the south.

Following the Battle of Britain, the station turned to the offensive with 1 and 266 Squadrons forming a wing with 19 Squadron at Fowlmere. On 6 April 1942, the first Mosquito NF.IIs with 151 Squadron arrived with night operations commencing on 30 April and moving out to Colerne on 30 April 1943 when the threat of enemy attacks had receded. The USAAF arrived at nearby Kingscliffe in August 1943 with P-38s of the 20th FG, positioning the 55th Fighter Squadron at Wittering until April 1944, while the RAF used Wittering for a number of equipment operational trials. The final combat operations were with 68 Squadron who flew Mosquitos from 2 February 1945 on anti-air-launched V-1 patrols until departing for Coltishall on 27 February. On 21 March Training Command took over the station. Wittering had its first hard runway constructed in 1941, in an almost E-W line with no room for any cross runways. A very short distance to the west of the original Wittering was the grass airfield of Collyweston, and in 1945 it was decided to combine the two sites and extend the runway on to the neighbouring airfield. Fighter Command returned to Wittering on 1 May 1946 with Spitfires of 41 Squadron and Mosquito NF.30s of 219 Squadron. They were later joined by the Spitfires of 19 Squadron and 141 Squadron Mosquito NF.36s, the ultimate RAF night fighter of this type. The station returned to Training Command on

20 February 1948, staying until 6 April 1950 when it closed to flying in preparation for some major improvements to bring it up to modern bomber standards. The 9,052 feet runway was upgraded and a 'Gaydon' hangar erected forward of the other hangars. The station was transferred to 3 Group, Bomber Command in 1953 and on 7 November the Bomber Command Armament School received in great secrecy the RAF's first "Blue Danube" atomic bomb. On 6 August 61 Squadron arrived with Lincolns, followed soon after by more of the same with 49 Squadron, although most of the time the squadrons were detached to Africa.

The jet age arrived in the form of Canberra B.2s when 76 Squadron re-formed on 9 December 1953, to be joined by 40 Squadron from Coningsby on 25 February 1954, 100 Squadron on 3 April and 61 Squadron re-formed in August making up briefly the Wittering Canberra Wing. With preparations planned to convert the station to a V-bomber base, the Canberras began to depart with 61 Squadron going to Upwood on 3 June 1955, on 15 November 76 Squadron went to Weston Zoyland to re-equip with Canberra B.6s to be used for air sampling following the nuclear tests at Christmas Island. No. 40 Squadron moved to Upwood on 1 November 1956, leaving 100 Squadron in residence until it disbanded on 1 September 1959.

Wyton, Cambs;
TL285741, 4 miles NE of Huntingdon alongside the A141.

Wyton was used as an RFC training base from 1 October 1916 until it closed on 2 September 1918. The site was selected to be an RAF bomber station as part of the RAF Expansion Scheme and work started at the end of 1935 with construction on a diamond-shaped site, with the landing ground to the north and the technical site to the south alongside the B.1090 road. Three 'C' Type hangars were constructed in a semicircle with a fourth located behind the easterly example. The substantial technical site was constructed behind the hangars to the standard Bomber Command pattern which made most of the bases so similar that when being posted from one to another, it was always possible to find your way around. Even though the construction was not completed, 139 Squadron re-formed on 3 September 1936 with Hinds, which had to be kept in the open until No.1 Hangar was completed in November. On 19 March 1937 Wyton introduced the Blenheim, with 139 Squadron making the first Bomber Command operational sortie of the war to graph the German fleet at Schillig Roads. On 1 November 1940 Wyton was transferred to 3 Group and the squadrons re-equipped with Wellingtons, but these were unsuitable for the strategic night offensive and had moved out by February 1941 to make way for the entry into service of the Stirling four-engine heavy bomber. This was introduced by 15 Squadron in March 1941, the first operation being to Berlin on 10 April. With the birth of 3 Group Pathfinder Force, the Stirlings departed to Bourn on 15 August 1942, and three runways were constructed, linked by a perimeter track with 36 heavy bomber hardstandings ready for the Lancasters of 83 Squadron until they departed for Coningsby on 20 April 1943. Mosquito FB.VIs of 109 Squadron had arrived from Stradishall in early September 1942 to be used to develop target marking by the use of Oboe blind bombing aid for the Pathfinder Force, and was replaced by 139 Squadron in June 1943 until it departed in February 1944. On 15 September

In its later use of the Lincoln B.2s, Wyton-based 44 Squadron had the upper surfaces painted white as on RF388 KM-C:44 Squadron. The squadron left Wyton for Marham on 7 February 1951 to convert to Washingtons. *RAF Museum P.016434*

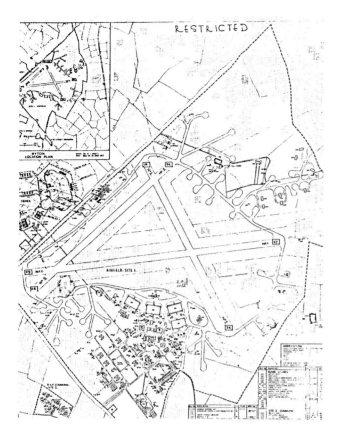

Left: The station was redeveloped from late 1935 to RAF Expansion Scheme standards including four 'C' Type hangars on a technical site on the south side of the airfield. Flying restarted from the incomplete site in September 1936 and from August 1942 three runways were constructed with a perimeter track containing 36 heavy bomber hardstandings dispersed around the site. In December 1946, Lincoln B.2s began to replace the Lancasters and in 1952, the station was selected for the strategic reconnaissance role with the main east-west runway being extended to the east to over 9,000 feet. The 1946 plan shows the station in its World War 2 configuration, prior to the Cold War developments. *RAF Museum*

1944 128 Squadron re-formed at Wyton with 20 Mosquito B.XVIs as part of the Light Night Strike Force, to be joined by 163 Squadron on 25 January 1945 until disbandment in August.

After a short period of relative inactivity, heavy bombers returned to Wyton with 15 Squadron on 19 August 1946 and 44 Squadron on 29 August both with Lancasters. They were joined by 90 Squadron on 11 November and 138 Squadron the next day. Lincoln B.2s began to replace the Lancasters with 44 Squadron in December 1946 with the other squadrons to follow. The squadrons flew on overseas exercises Sunray and Sunbronze, followed by anti-shipping trials in October 1947, and 15 Squadron took part in Exercise Ruby, which involved dropping 12,000lb Tallboy bombs on redundant German concrete coastal structures. By December 1947 the total maximum establishment of the four squadrons was six aircraft each, although only 18 were on the base. With the withdrawal of the Lincolns to be replaced by Washingtons, 90 and 138 Squadrons disbanded on 1 September 1950, 15 Squadron departed for Marham on 29 November 1950 and 44 Squadron left for Marham on 7 February 1951. This allowed the arrival of dispersed USAF B-50s and B-29s of the 93rd BG and 2nd BG, the Americans departing on 16 May 1951. In 1952 Wyton was selected as a strategic reconnaissance station replacing Benson, with considerable improvements to the base including extending the main runway to over 9,000 feet the east across the Ramsey to St Ives road to allow the operation of V-bombers.

Left: No 44 Squadron operated Lincoln B.2s at Wyton for various periods from October 1945 to January 1951. The squadron was based at Wyton from 29 August 1946 to 7 February 1951. Lincoln B.2 RF423 KM-K:44 Squadron was one of the earlier aircraft with the squadron and still had the World War 2 camouflage finish. *RAF Museum* P.016435

Chapter 5

Britain's USAF Airfields

Following the withdrawal from Europe, particularly Britain after the end of World War 2 with the building tension of the Cold War, the USAF returned to establish a number of major bases in the southeast quarter of Britain. This allowed the USAF to deploy strategic bombers capable of delivering a nuclear attack on Soviet Russia, as well as strike fighters, strategic reconnaissance aircraft and a fleet of transport and air-to-air refuelling aircraft. At the height of the Cold War some 14 airfields were active in Britain with many more in continental Europe.

Alconbury, Cambs;
TL295795, four miles NW of Huntingdon.

A large grass meadow at Alconbury was acquired in 1938 for use as basic dispersal site for the use of No.1 Group Advanced Air Striking Force which, in the event of war would occupy rudimentary airfields in France for short-range bombing of targets in Germany, while also providing support for the British Expeditionary Force (BEF) by attacking battlefield targets. On 17-18 May the Battles of 63 Squadron, based at Upwood, arrived at the site with the air and ground crews accommodated under canvas, testing for the first time the principle of Satellite Landing Grounds (SLG) and dispersal system, with other sites chosen for other stations. With the outbreak of World War 2, Alconbury became more active as the SLG for Wyton and XV Squadron with Blenheims took up residence on 14 April 1940, the personnel returning every night to Wyton. The first active operation was on 10 May when two Blenheims were sent on reconnaissance duties to determine the status of river bridges in the Low Countries, leading to an unsuccessful raid on an airfield south of Rotterdam, with the loss of half the squadron. The range of the Blenheims was too short to reach German targets, and they were replaced by Wellingtons, the first action by XV Squadron from Alconbury being on 22 December 1940. In February 1941 XV Squadron was replaced by 40 Squadron, and moved back to Wyton to become the second Stirling unit.

The rudimentary facilities at Alconbury were dramatically improved in early 1941 with the construction of runways to the standard bomber pattern and a connecting perimeter track and 50 loop-type dispersals. Alconbury was still a Wyton satellite, but was capable of operating the Stirlings of XV Squadron. Two T.2 hangars were erected, and the remaining domestic buildings were rather basic and dispersed outside the airfield, since more permanent accommodation existed at Wyton. RAF bomber operations continued until August 1942, when 156 Squadron moved to Wyton and then Warboys as Alconbury had been allocated to the Americans.

The B-24D Liberators of the 93rd Bomb Group, US 8th AAF arrived on 7 September 1942 and within three days 34 aircraft were on the airfield. After a month of intensive training and preparation, the 93rd BG went into action on 8 October on the hazardous daylight bombing raids against strategic targets in Germany and enemy occupied Europe. When the 93rd Bomb Group left in December 1942, they were replaced by the 92nd Bomb Group with B-17s ready for daylight operations to continue on 1 May 1943 until March 1944. Then the 482nd Bomb Group was tasked with the training of 8th AF pathfinders and the operational development of radar equipment including H2S. In support of D-Day, the 482nd BG sent 49 B-17s on attacks, disrupting surface traffic behind the beachhead after which bombing operations slowed down until inactivation on 31 October 1945, leading to the station being put on C&M on 30 September 1948.

With the increased tensions of the Cold War, interest in Alconbury by the USAF returned, but on 24 August 1951 it was designated a 'USAF Primary Installation' with a request for a 11,000ft runway, but due to cost restrictions, a 9,000ft

Right: A view from the Alconbury Air Traffic Control Tower of the RF-4Cs of the 10th TRW in August 1966. *Michael Bowyer*

By October 1972 when this RF-4C AF64-019 of the 10thTRW was seen at Alconbury, the aircraft carried base identification 'AR' in white on the fin.
Author

runway was approved, together with a major modernisation over a two-year period. The planned role was as a USAF Air Depot with the provision that if war was declared, the F-84s would move in from Manston. On 1 June 1953 the Americans returned in the support role, and in 1954 asphalt overruns were laid at the end of runway 12/30 to bring the overall length to 10,046 feet. In 1955 Alconbury was designated a satellite to Sculthorpe, from where on 15 September the B-45A jet bombers of the 86th BS, 47th BW arrived. These were replaced by B-66Bs in May 1958 until departure on 5 August 1959 following the French Government requesting the departure of all American military from their territory. As a result the RB-66s of the 1st and 30th squadrons of the 10th Tactical Reconnaissance Wing (TRW) arrived from Spangdahlem on 25 August 1959.

From 12 May to August 1965 there was a dramatic change for the 10th TRW with the arrival of the RF-4C Phantom IIs, initially re-equipping 1st TRS, followed in 1966 by the 30th TRS. These were joined on 15 August 1966 by the RF-4Cs of the 32nd TRS, 26 TRW from Toul/Rosieres, with 18 aircraft in each squadron. These aircraft were the first all-weather, day and night tactical reconnaissance aircraft in USAF service. With a crew of two, a pilot and systems operator, the aircraft was powered by two GE J79 turbojets which developed over 34,000lb of thrust with reheat. These engines gave a top speed of 1,600 mph and an altitude of 50,000 feet could be reached in just over six minutes. For low speed control on the approach and landing high lift devises and boundary control allowed speeds of around 150 mph. The reconnaissance versions of the Phantom had three camera positions in the nose for forward, sideways and vertical panoramic graphy, and because they worked at a very high shutter speed, highly detailed sharp images could be obtained in high-speed, low-altitude flight. For operations at night an infra-red image system was carried. Later reconnaissance versions were equipped with an advanced Side Looking Airborne Radar (SLAR) to achieve Tactical Electronic Reconnaissance (TEREC). The SLAR captured high resolution radar images, and TEREC could locate enemy transmitters. When data-link was later installed, information from both systems could be transmitted to ground units giving up to date information almost in real-time. From January 1970, the Phantoms were camouflaged and from the autumn of 1972 the letters 'AR' were on the fins. With the change from -dedicated aircraft to detachable reconnaissance pods, the Phantoms of 32nd and 30th were withdrawn by 28 February 1976.

In their place came the F-5E Tiger II-equipped 527th Tactical Fighter Training Aggressor Squadron, arriving on 1 April 1976 to provide dissimilar air combat training, simulating Soviet combat aircraft. The aircraft were shipped in by C-5As, the first of 20 arriving on 21 May 1976, but it was 1 January 1977 before the unit was fully operational. The F-5E was an advanced version of the Northrop F-5 Freedom Fighter used widely by NATO air forces. It was powered by two GE J-85 turbojets developing 5,000lb thrust with reheat, which gave a top speed of around 1,000 mph and a ceiling of up to 50,000 feet. Armament consisted of two 22mm cannons in the nose with two AIM-9 Sidewinder air-to-air missiles carried on wingtip launch rails. In addition up to 7,000lb of weapons could be carried under the wings and fuselage. With its compact dimensions of 26ft wingspan and length of just over 48 feet, the F-5E was similar in size and radar signature to the MiG-21, which was the most common Warsaw Pact tactical aircraft. To make the training even more realistic, the F-5Es were painted in Warsaw Pact camouflage schemes.

In April 1983 the unit was re-designated the 527th Aggressor Squadron (AS) with detachments from NATO and USA-based squadrons being trained in mock combat. The 527th AS had three main tasks; responsibility for Dissimilar Air Combat Training (DACT); create a training programme using

Left: The accommodation and facilities at Alconbury were fairly basic, including the hangarage as shown behind Vulcan B.2 XM656 of the Waddington Wing, which visited in October 1972. *Author*

the latest Soviet combat methods; and to support British air defence in times of tension. The Soviet combat methods were based on secret information, later found to have been achieved using a classified special squadron of Soviet fighter aircraft acquired by various means. The aggressor unit operated from Alconbury for 12 years honing the interception skills of allied pilots, but when the more advanced MiG-29 Fulcrum came into service, the F-5Es were replaced by F-16s in the aggressor role. Meanwhile an extensive area of hardened aircraft shelters had been built on the eastern side of the airfield, to protect the remaining RF-4 Phantom fleet.

On 12 February 1983 a new shape arrived at Alconbury in the form of the ultra-high-flying all-black TR-1A battlefield-surveillance version of the U-2, 15 of which equipped the 95th Recon Squadron, 17th Recon Wing which had formed on 1 October 1982. To accommodate these specialist aircraft, five prefabricated ready sheds were erected, 13 extra-wide hardened aircraft shelters and aprons were built, and the north side perimeter track was widened. Due to the demanding nature of landing the aircraft in a crosswind, the remaining short cross runway was resurfaced. The TR-1s were capable of very quiet operation up to 90,000ft altitude, equipped with highly sophisticated radar and passive sensors giving high-resolution images in all weathers round the clock while remaining at a safe distance from interception. By 1990 the 95th RS had reduced its compliment to twelve aircraft which were used for SIGINT and high-resolution -reconnaissance missions, the aircraft being fitted with ASARS-2 (Advanced Synthetic Aperture Radar System), which is a very accurate sideways-looking radar system and when flying at 65,000 feet at the edge of the battle area could see targets some 30 miles inside enemy territory. Whereas satellite images gather strategic information, systems such as the TR-1A were able to gather high-quality images of tactical situations such as armoured formations, mobile command posts, troop dispositions and airfield situations and transmit the information in real time via a classified data link to a Central Processing Station (CPS). Here the information was evaluated and passed on to allied troops in the field, who could prepare for an attack

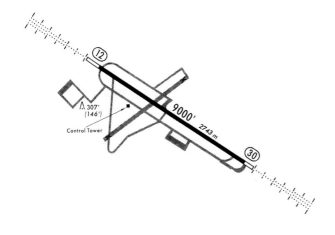

Left: Alconbury first opened to flying in May 1938 as a SLG for nearby Wyton. The site was developed in 1941 with the construction of three runways in the standard bomber pattern with perimeter track giving access to 50 dispersals. The technical site was on the southwest side of the airfield with two T.2 hangars and very basic accommodation, as the personnel were based at Wyton. The USAAF arrived in September 1942 with B-17s when the facilities were improved, and departed in October 1945, when the station was put under C&M. The Americans returned in 1951 when the main runway was extended to 9,000 feet and further improvements made to the accommodation. In 1954 the runway was further lengthened with tarmac over-runs at each end to a total length of 10,046 feet as a satellite to Sculthorpe when B-45s and later B-66s arrived. Alconbury was a major USAFE RF-4C Phantom base from May 1965 until August 1987 during which time HASs were erected on the northeast boundary. Following operations by TR-1As, and then A-10As, the airfield closed to flying in September 1995. Although the airfield remains, the north end is a commercial park and the USAFE retain a domestic presence on the southern boundary with a plastic replica of an F-5E by the main gate. The chart shows the runway layout in August 1964 with the two shorter runways disused and an operating apron on the south side and an additional dispersal on the northwest corner. *Jeppesen*

well ahead of the enemy being in a position to attack. The Precision Locator Strike System (PLSS) goes a step further by sending the information to a ground station or AWACS (Airborne Warning and Control System) from where strike aircraft can be directed to the designated targets. From a safe distance the aircraft can then launch fire-and-forget weapons which will be automatically directed to the targets using the PLSS data. The PLSS missions involved three TR-1As at a time, each flying a race-track pattern close to the battle area. When a hostile radar signal was detected, each TR-1A passed the information to an AWACS or CPS, and by complicated triangulation, the target location could be determined, and the position passed to the strike aircraft to make a hit. Even if the enemy radar was switched off, the location was received in one pulse and the position marked. Operating at altitudes of around 65,000 feet, the TR-1As could see some 200 miles into hostile territory.

On 26 August 1987 the last sortie was flown by a 1st TRS Phantom and three days later the squadron disbanded, as on 20 August the 10th TRW was redesignated the 10th Tactical Fighter Wing with two squadrons totalling 36 A-10As moving in from Bentwaters. The first aircraft did not arrive until 4 January 1988, followed by two more on 15 April as part of the 509th Tactical Fighter Squadron (TFS), from the 81st TFW. There were enough A-10As tank busters to activate two TFWs, and the 511st TFS began to arrive at Alconbury on 1 July when the units were formally activated with the 36th A-10 completing the changeover on 15 September. Meanwhile the F-5s of the 527th AS ceased flying on 22 June and on 14 July the squadron moved to Bentwaters, leaving its aircraft behind for disposal overseas.

In late 1990 three TR-1A were deployed to Taif AB in Saudi Arabia to operate in Desert Storm, the first Gulf War. On 27 December 23 A-10s of the 511th TFS were deployed to Damman/King Fahd International Airport, where they were absorbed into the Myrtle Beach-based 354th TFW. After the end of the Gulf War, the 511th TFS returned to Alconbury on 8 June 1991, followed soon after by the TR-1As, the 95th RS being disbanded on 15 September 1993, to be reactivated at Mildenhall as the 95 RS on 1 July 1994. With the threat of the Cold War melted away, the A-10s starting in October 1991 to join the Wing at Myrtle Beach, the last departing Alconbury on 30 March 1992, leaving four TR-1s as the sole residents until their departure to Mildenhall.

Although the Cold War was no longer a threat, Alconbury continued as a USAFE air base with the arrival of Special Operations Hercules and MH-53 helicopters. On 1 September 1993, the special-operations units began to depart and on 31 March 1995 the airfield closed to flying and returned to the MOD on 30 September. A USAFE presence is still at Alconbury in the SW corner as a domestic site for the staff working at the Joint Intelligence Centre at Molesworth and the north end of the site for commercial development, while the airfield facilities remain with the long runway.

Bassingbourn, Cambs;
TL335458, 4 miles N of Royston, Herts by the A1198.

Bassingbourn is probably most famous as the 8th USAAF base from where the B-17 *Memphis Belle* made its 25th combat sortie, recorded for prosperity in the film made by William Wiler and his team. Claimed to be the first to achieve this milestone, it was in fact the second, the premier place going to the "Hell's Angels", a B-17 operating from Molesworth, but the film team were already waiting at Bassingbourn ready to rewrite history. The other main memory of Bassingbourn is as the RAF Canberra jet bomber Operational Conversion Unit (OCU), but the station did play a modest role in the Cold War.

Bassingbourn opened in March 1938 as probably one of the last of the RAF Expansion Scheme stations with a semicircle of four 'C' Type hangars with a watch office in the centre. The Station HQ was located inside the main gate with the guardroom on the right, and the technical site included the usual workshops, stores, water tower and barracks, with officers' mess and married quarters to the south and additional housing on the opposite of the A1198. The first units to arrive from 2 May 1938 were 104 and 108 Squadron with Hinds, soon replaced by Blenheims, with 104 Squadron from 10 May and 108 Squadron by 24 June 1938. In May 23 Ansons were added to the Blenheims to provide the RAF Volunteer Reserve aircrew with experience in operational aircraft, the role of the station being to train the crews for 2 Group Blenheim squadrons. In September 1939 both Blenheim squadrons moved to Bicester and Wellingtons arrived with 215 Squadron on 24 September to train the crews of 3 Group, with operational training continuing until February 1943.

Much of the bomber training was dispersed to nearby Steeple Morden when runway construction commenced in December to the usual three-runway pattern with the main one 6,000ft long. Around the perimeter was a concrete track with 54 bomber dispersals. The main runway reopened on 24 April 1942 ready for the arrival on 14 October of the first B-17s in Britain, operated by 91st BG when the station was handed over to the USAAF. After a period of local training the first combat mission was flown by 34 B-17s of the group on U-boat pens at Brest. Daylight operations continued against various targets in France until the first raid over Germany on 27 January 1943 when the navy installations at Wilhelmshaven were bombed. The strength increased to 46 B-17s by 13 May with sorties continuing over Europe, including Berlin, with growing losses to the crews. Probably the worst was on Schweinfurt on 17 August 1943 when ten of the B-17s of the 91st BG failed to return with the loss of 97 men, and of those that returned 11 aircraft were damaged, one of which was beyond repair. Only three B-17s were at dispersals the next day available for operations, but the strength of the 91st was restored by 14 October. Hazardous day bombing operations continued until the final combat mission to Pilsen on 25 April 1945,

Left: Canberra PR.3 WE135 of 231 OCU outside one of the 'C' Type hangars at Bassingbourn after the departure of the USAF B-29s and B-50s. *Author*

having completed 9,591 sorties from which 197 aircraft did not return with loss of life at over 600 young Americans. The B-17s began to depart on 27 May with the last Americans leaving on 23 June 1945 and three days later the RAF returned under the control of 47 Group Transport Command, ready for the arrival in August of the Liberators of 102 Squadron.

RAF Transport operations continued until the end of October 1950, when due to the Korean War, the Americans returned on 25 August with 15 B-29s of the 353rd BS, 301st BG and further B-29s of the 353rd BS were also dispersed from Lakenheath. As the main runway at Bassingbourn was too short for the operation of fully laden bombers, work began in March 1951 on improvements to the main runway and improving the facilities to be able to accommodate two B-29/50 Squadrons. The first rotation from January 1951 was ten secret RB-50Bs of the 38th SRS 55th Strategic Reconnaissance Wing which stayed until May 1951, the role of the aircraft being to patrol around the periphery of Eastern Europe to produce radar maps. Also arrived in January were 15 B-50Ds of the 341st BS, 97th BG, the crews practising raids over Europe and interceptions with RAF fighters. Rotations continued until September 1951 and the Americans left Bassingbourn by 1 April 1953, although there had been plans to extend the runway to 9,000 feet, but the cost was not worth the short remaining service of the B-29/50s, thus ending this short period of Cold War activity.

The RAF returned in the autumn of 1951 and preparations were made to accept the training role for the conversion of crews to the Canberra jet bomber. Initially in late 1951 Mosquito and Meteors arrived and on 1 December 231 OCU re-formed with the responsibility of training Mosquito PR aircrews, but on 14 January 1952 the first Canberra B.2 arrived, and as more aircraft were delivered the first course commenced on 27 May 1952. The full strength of 26 Canberras was reached after a year and included the first dual-control T.4 and seven PR.3s from November 1953. To cope with jet operations improvements included a new control tower on the south side, a large aircraft servicing platform in front of the hangars and additional taxiways. Not just conversion training was provided, but also operational training including LABS (Low Approach Bombing System) where tactical nuclear weapons could be tossed as the aircraft climbed, and after release departed from off the top to avoid the nuclear blast effects. Jet bomber training continued for the RAF, RN and overseas air forces pilots and navigators, but with the reduction in RAF Canberra squadrons and on 19 May 1969 16 Canberras departed for Cottesmore. On 29 August 1969 the RAF ensign was lowered for the last time and the base taken over as an Army training establishment with the Queen's Division, a role which continues. Most of the airfield has been converted to a training area with a golf course alongside the A1198, but two short sections of the runways remain, and three of the 'C' Type hangars, the furthest one from the road having been demolished.

Bentwaters, Suffolk;
TM350530, by A1152, 5 miles NE of Woodbridge.

The construction of Bentwaters began in late 1942 as a temporary wartime airfield to house a USAAF bomb group, the airfield being named after a farmhouse located where the main runway was built. The incomplete base was opened on 17 April 1944 under C&M controlled by RAF Bomber Command, as it was surplus to American needs. When the construction was almost complete, work was stopped in May 1944, although it did attract a number of emergency landings, and the runways being made safe for operations by 10 October. With its advanced position near the coast, the airfield was ideally situated for the use of escort fighters protecting the heavy bombers, resulting in the airfield being declared open on 22 November 1944 in the North Weald sector of 11 Group, becoming the last new airfield to be commissioned for operational purposes. On 11 December RAF Mustangs of 129 Squadron arrived, to be joined by five other squadrons escorting RAF Lancaster and Halifax daylight raids until the final operation on 4 May 1945.

With the three concrete runways, excellent approaches and a pair of T.2 hangars, Bentwaters was ideal for jet fighter operations and on 5 October 1945 sixteen Meteor F.IIIs of 124 Squadron arrived from Boxted and on 4 April 1946 it was renumbered 56 Squadron. On 2 June 1946 74 and 245 Squadron arrived with Meteor F.IIIs to form the first Meteor

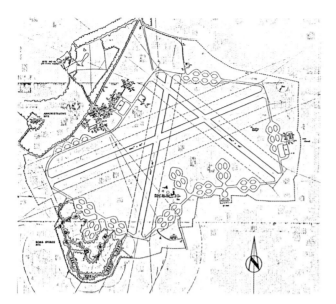

Left: The airfield at Bentwaters was one of the last stations to be constructed during World War 2 and consisted of the standard layout of three runways with dispersals around the perimeter track including one area surrounded by trees on the southwest corner where munitions were stored. In 1950 the technical site was very basic with one T.2 hangar and another on the eastern side of the airfield. The main NE-SW runway was later extended and the accommodation site was located off the station on the northwest side and considerably developed by the USAF, who took over the station in March 1951 ready for the first F-86 Sabres of the 81st TFW to arrive on 5 September 1951. The last USAFE A-10A left in May 1993 when the base closed. *RAF Museum*

Fighter Wing, followed soon after by the second Meteor wing at Boxted. On 9 June both 74 and 245 Squadrons departed for Colerne after a very short stay, followed by 56 Squadron to Boxted on 16 September, Bentwaters then taking on the Meteor training role when 226 Operational Conversion Unit (OCU) arrived from Molesworth on 10 October. With Bomber Command abandoning nearby Woodbridge, it came under the care of Bentwaters, although the latter was inactive from 1 September 1949 until 1 July 1950 when it was put under C&M.

With the increasing Cold War tensions, the US government was seeking bases in East Anglia, and was offered Bentwaters as a base for fighters to escort bomber formations. On 16 March 1951 the USAF arrived to prepare the base for 25 F-86A Sabres, the first arriving on 5 September as part of the 81st Fighter Interceptor Wing, the first two squadrons making Shepherd's Grove their base, plus 91st Squadron which arrived at Bentwaters on 26 September. The 91st FIS re-equipped with F-86F Sabres, but moved out in the autumn of 1953 to allow improvements, including the lengthening of the main runway with gravel overshoots at each end and a runway arrested barrier at the western end. Nuclear weapons shelters were erected on the SE side of the airfield, the main technical site being on the north side. With the improvements complete the redesignated 81st Fighter Bomber Wing arrived on 1 April 1954 equipped with F-84F Thunderstreaks and was fully equipped by early 1955. On 13 December 1954 the F-86D Sabre night fighters arrived from Manston with the 87th FIS armed with Mighty Mouse rocket pods, as part of the 406th FIW, and remained until 8 September 1955.

On 30 April 1958 the 92nd FBS arrived from Manston, taking up residence on the east side of the base, with the 91st dispersals on the opposite side. The base became the F-84F-equipped 81st Tactical Fighter Wing (TFW) in a change of title on 8 July, becoming twinned with Woodbridge once more. New equipment began to arrive on 10 August 1958 in the form of the first seven F-101C Voodoos of the 27th FBW, having been flight refuelled by USAF KC-135 tankers on the delivery flight. On 4 December the first five Voodoos arrived for the 81st FBS, the 41 F-84Fs being transferred to the West German Air Force. The Voodoo was a large twin-engine fighter bomber with a nuclear weapons capability, the 91st and 92nd TFS receiving the new type between December 1958 and March 1959. The sight and sound of these aircraft became familiar in the surrounding area as they built up hours in operational training, including LABS manoeuvres, where the aircraft climbed steeply from low level into a loop, releasing the weapon in the climb, in effect toss-bombing the target. As the aircraft reached the top, the pilot rolled the aircraft level, and departed rapidly to be as far away as possible from the explosion. To reach targets in Eastern Europe would have meant in effect a one-way sortie. The F-101A began to be replaced by the F-4C Phantom II, the Voodoos beginning to depart in July 1965, with the first

Below: The 81st FBW arrived at Bentwaters in April 1954 with F-84F Thunderstreaks, and upon replacement by Voodoos the F-84Fs were passed to the West German AF by December 1958. F-84F 26713 was with the 81st FBW during the time at Bentwaters. *Michael Bowyer*

Above: At the Open House at Bentwaters on 26 May 1973 C-141 67956 of 436 MAW was in the static park, and probably had brought in supplies for the visitors. *Author*

Phantoms for the 81st arriving on 4 October 1965, and the conversion completed by 26 April 1966. The F-4C was armed with an AIM-7 radar-guided air-to-air missile and in addition a M61 22mm Vulcan cannon in a pod mounted under the centre fuselage. This version of the Phantom was developed for close air support and ground attack with a load-carrying capacity of up to 16,000lb of tactical nuclear or conventional weapons. The last F-101s left Bentwaters on 3 January 1966. Between August and October 1973 the 91st and 92nd TFS converted to the F-4Ds with improved weapons guidance and the capability of releasing laser-guided smart weapons. The main runway had by this time been extended to 10,900 feet with a warm-up pad at the eastern end, and the other two runways had been disused. Hardened Aircraft Shelters (HAS) had been erected to protect the based aircraft against anything apart from a direct hit and more hangars were also built.

The replacement of the Phantoms at Bentwaters began on 24 August 1978 when the first three A-10A Thunderbolt tank busters arrived for the training of ground crews, with further A-10As delivered from December, allowing the 92nd TFS to start rearming. The A-10A, while not being a very attractive aircraft, and known as the Warthog, was a flying anti-armour gun which could fire its 1,350 rounds at a rate of 4,200 rounds per minute from the nose-mounted revolving seven-barrel 30mm cannon armour-piercing shells, while also dropping bombs. In addition to the cannon, the Warthog can also carry up to 16,000lb of weapons including AGM-65 Maverick TV guided air-to-ground weapons and a laser system pod for target designation. When flying close air support (CAS) missions, the aircraft were flown in pairs two or three miles apart at very low levels to avoid detection of hostile radars and was very manoeuvrable at low speeds and heights above the terrain. The combat radius of the A-10 was in the region of 250 miles, which was sufficient to reach targets in East Germany from a Forward Operating Location (FOL). The combat endurance of the aircraft was up to two hours, and the 30mm cannon had enough ammunition to make up to 15 attacks at ground targets. The aircraft was capable of operating from short unprepared strips up to 1,500 feet long as well as sections of roads. With a turning circle of 4,000 feet and the gunsight calibrated at 4,000 feet allowed the pilot to attack a target without having to fly over it, a one-second burst firing 70 rounds at a target. With a 360-degree turn taking only 16 seconds, a pair of A-10s could bring continual fire to bear. The pilot was protected by a titanium bathtub' capable of defending direct hits from the ground and the fuel tanks were protected with an explosion-resistant foam layer to suppress fire in the event of a direct hit. It was a very robust aircraft, capable of taking great punishment and still bringing its pilot home. It could still fly with half a tail shot as well as one engine, the latter shielded by the straight wing to reduce infra-red heat emissions. So many A-10As were deployed to Bentwaters that the 91st transferred to Woodbridge, and three more fighter squadrons formed, the 509th and 510th both activated on 1 October 1979, and were followed by the 511st on 1 January 1980, bringing the strength of the 81st TFW to an unusual six squadrons spread over the two bases. There was a rumour at one time that Bentwaters and Woodbridge may have been connected by a single long runway, no doubt suitable for a Shuttle landing if it ever came near Britain. Once the 81st TFW was fully established, in addition to low-level training around the UK, 40 percent of the total strength was detached to FOLs in Germany, where there

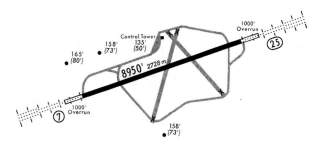

Left: The construction of Bentwaters started in late 1942 as a temporary USAAF bomber base and was opened in November 1944 with three runways and a rudimentary technical site on the northern boundary including two T.2 hangars. RAF Mustangs arrived the following month and stayed until May 1945. With its excellent approaches, Bentwaters became home of the first Meteor jet-fighter wing from October 1945. The USAF took over in March 1951 with F-86 Sabres arriving six months later. The main runway was extended to 8,950ft in 1953, and the other two runways became disused, with F-101 Voodoos based there from August 1958 until January 1966. With the arrival of Phantoms in October 1965, the main runway was further extended to 10,900ft and HASs were erected. The January 1969 approach chart shows the original extension of the main runway with 1,000ft overruns at each end. A-10As began to replace the Phantoms in August 1978, and the base closed to flying in May 1993. *Jeppesen*

would be eight aircraft weekly at each base. By the early 1980s the A-10s occupied the HAS on the south side of Bentwaters where additional dispersals had been constructed in a wooded area in the SW corner.

With the phasing out of the 10th TRW -reconnaissance Phantoms at Alconbury, the A-10As of 509th TFS transferred to Alconbury on 15 April 1988, followed by the 511st on 1 July. They replaced the F-5Es, by then at Bentwaters, of the 527th Aggressor Squadron, which at the end of the year began to re-equip with F-16Cs. The aggressor training was mainly for the benefit of the USAFE F-15 squadrons, but the F-16 equipped air forces from Belgium, Holland and Norway also took advantage of the training which focussed on the dissimilar combat situation which could simulate an attack by multiple aggressors against two F-15s. This provided the pilots with the skills required to survive an attack by multiple enemy aircraft, as the Warsaw Pact fighter strength was double that of NATO. All the exercises were recorded to allow an analysis of each pilot's performance during a subsequent debriefing, and it was proved that after ten DACT missions, the pilot's chance of surviving in actual combat were doubled. Because of limitations on the British combat ranges, due to weather and restrictions on supersonic flight over land, the DACT training programmes were moved to Decimomannu in Sardinia where an Advanced Combat Manoeuvring Instrument (ACMI) range was set up from 1978, allowing realistic supersonic air battle training to be conducted without restrictions.

In late 1989 the F-16s began to depart Bentwaters, although the 527th AS was not formally deactivated until 30 September 1990. With the end of the Cold War, the base began to run down rapidly, with the 510th TFS going to Germany on 4 January 1993. The 92 TFS was inactivated on 31 March and the 81st stayed on until 21 May when the last two A-10As departed as the last USAFE aircraft to leave the base. Meanwhile during the A-10A deployment with so many aircraft, the domestic aspects of the base had developed considerably. The large domestic site on the northwest side of the airfield was developed across the A1152 with a hospital, large shopping mall, theatre and a variety of married quarters, some of the latter now privately owned.

The airfield was used for two weeks in September 1994 as a Harrier deployment exercise within the wooded area on the south side of the base. The airfield is now privately owned by Bentwaters Parks, the control tower being the owner's offices, but its remote location makes the site generally unsuitable for development. The Cold War Museum opened on 27 May 2007 in the Wing Command Post building and is slowly adding to their aircraft collection, with regular open days organised.

Left: When first arrived at Bentwaters the A-10As of the 81st TFW were painted in a two-tone grey camouflage as shown on AF77 237 visiting Abingdon in September 1979. *Author*

Brize Norton, Oxfordshire;
SP295060, 6 miles SW of Witney

Brize Norton is one of the major RAF master bases, concentrating on air transport world wide and air-to-air flight refuelling. Although it is named after a small village to the NE, the station itself has absorbed Carterton with its shopping and accommodation. Construction of the station commenced in 1935 along the lines of the RAF Expansion Scheme with the full compliment of five 'C' Type hangars in the technical area on the NW side of the flying field. By the end of 1944 there was a total of 44 assorted hangars around the airfield, including Type 'D' and 'E' and six concrete 'Lamella' storage hangars for 6 MU. The first flying occupants of the airfield was 2 FTS which moved in before completion on 13 August 1937, and by the start of World War 2 the airfield was busy with flying training and the work of the MU. Two of the 'C' Type hangars and other buildings were destroyed in a raid by a pair of Ju.88s on 16 August 1940 with 46 training aircraft destroyed, although fortunately only one civilian was killed.

On 15 July 1942 the role of Brize Norton changed from training to transport with the arrival of the Horsa troop-carrying gliders of the Heavy Glider Conversion Unit (HGCU) to train the pilots for the airborne forces. No.60 MU was also responsible for the assembly and despatch of many Horsas, the major assemblies being delivered from the woodworking industry around the country. With a strength of 56 Horsas, there were also 34 Whitley glider tugs. In April 1943 the two runways were resurfaced, followed by the main task of training for the D-Day landings in Normandy. After this the HGCU strength was reduced and moved to North Luffenham in March 1944. Brize Norton then became a 38 Group operational station from 13 March 1944 with the Albemarles of 296 and 297 Squadrons used for dropping members of the 5th Parachute Brigade in Normandy on the night of 5/6 June, who prepared a Horsa landing ground by the River Orne ready for 17 Horsas later in the day. The 6th Airborne Division was taken from Brize Norton in 40 Horsas in the early evening of 6 June, and the Albemarles continued to support the advancing armies with supply drops, while still maintaining their glider towing skills, which were next used in Operation Market Garden, the attempt to capture the bridge across the Rhine at Arnhem. The rather unsuitable Albemarles were soon replaced by four-engine Stirlings, which left for Earl's Colne on 29 and 30 September 1944. With the loss of so many glider pilots at Arnhem, there was a need to train more pilots for the planned airborne Rhine crossing, resulting in the Whitleys and Horsas returning from North Luffenham on 15 October 1944, and as there was more than one training unit, the one at Brize Norton became 21 HGCU with American Waco Hadrian gliders joining the strength. There were enough glider pilots trained by February 1945, but training continued to provide for a possible requirement in Asia.

As soon as the war was over in Europe, 6 MU became busy dealing with some of the 200 or so captured enemy aircraft, which were evaluated at Farnborough, the first arrival being a Ju.88 on 10 May 1945, with many more diverse types to follow. With the departure of the Horsas at the end of 1945, Transport Command took over the station where tactical airborne delivery methods were investigated by the Transport Command Development Unit (TCDU) until it left for

Left: Construction of the RAF Expansion Scheme station at Brize Norton started in 1935 and was completed by August 1937. It included five 'C' Type hangars as part of the technical site located on the western boundary of the airfield. The station was initially used for flying training until 1942, and then became responsible for heavy glider training from July 1942 until March 1944 when preparation were made for the D-Day landing support. Two of the 'C' Type hangars were destroyed by enemy action in August 1940, and the two runways were resurfaced in April 1943. After D-Day and Arnhem heavy glider training returned in October to make up for the losses of pilots and continued until the end of 1945. The airfield was allocated to the USAF from June 1950 and amongst the many improvements made was the extension of the main runway to 9,000 feet and the widening and strengthening of the perimeter track with a number of dispersals. The first arrivals were the massive B-36s, followed by B29/50s, the jet B-47s and later B-52s, with support from flight refuelling tankers, such as the KC-135s. The airfield was returned to the RAF in April 1965, and Brize Norton became a major RAF strategic and flight refuelling base with Belfasts, VC-10s and TriStars, to which have now been added the C-17s with 99 Squadron. Future developments include the arrival of the A.300-based flight refuelling tanker/transports to replace the aged VC-10s and TriStars. *Pooley's Flight Guides*

Abingdon on 30 June 1949. On 4 July the station became under the control of Training Command with Harvards of the CFS Examining Wing with nearby Fairford as a relief landing ground and 60 MU was busy disposing of surplus aircraft, including Spitfires, although in January 1948 it did receive its first Meteors.

With the increased tension from the Soviets, the station was taken over on 1 June 1950 by Bomber Command on behalf of the American Strategic Air Command, ready for USAF B-29s to move closer to the front line. In addition to Brize Norton, Fairford, Greenham Common and Upper Heyford were selected in May 1950 for major developments including 9,000ft runways, wider perimeter tracks, strengthened dispersals and secure weapons storage. The American advance party arrived on 7 June 1950 ready for the official handover to the USAF on 16 April 1951. The first USAF aircraft to arrive on 27 June 1952 was a mixture of the enormous B-36Ds and B-36Fs from the 11th Bomb Wing from Carswell AFB, which stayed for a deployment of fifteen days, followed by regular rotations from the USA. The B-36 had originally been conceived with a capability of bombing Germany from bases in the USA. It was powered by six P&W R-4360 piston engines mounted on the trailing edge of the wings, the prototype making its first flight in August 1946, and entering service with SAC in 1948. The B-36 weighed 160 tons, which was double the weight of the B-50 and with a range of 8,000 miles could easily reach Moscow. The improved B-36D had additional GE J-47 jet engines in paired pods under the outer wings giving increased speed over the target area. This version also had snap-action bomb doors and a formidable defensive armament of 12 20mm cannons paired in six remotely-controlled turrets, plus a 37mm cannon in the nose. During their time in Britain these B-36Ds flew fighter affiliation exercises with Meteors and Vampires of Fighter Command.

In December the 301st BW B-29s were deployed, followed by the B-50As of the 43rd BW from March to May 1953. Preparations were then made for the arrival of the jet powered B-47Bs of the 305th BW from MacDill, two squadrons being in residence from 4 September to December 1953. The B-47 rotations continued from December 1953 to November 1954, and in December 1954 the first tanker unit arrived with KC-97Gs of the 321st Air Refuelling Squadron (ARS). Further tanker squadrons rotated until the September 1955, when repair work was started on the main runway. It reopened in July 1956 with the B-47Es rotating only one squadron at a time, and B-52B Stratofortress of the 93rd BW arrived on 16 January 1957, from Castle AFB and stayed for a few days, the first to visit Britain. The B-47E and KC-97 deployments continued from late 1957 into 1958, and the first KC-135A tankers to arrive in Britain landed at Brize Norton on 27 March 1958, from the 99th ARS at Westover, Maine.

From April 1958 there were reduced numbers of USAF bombers with six B-52Ds of the 92nd BW to participate in the annual Anglo-American bombing competition. There would from time to time be as few as nine B-47s at Brize Norton for a three-week stay and at various times there were also detachments of RB-47s, the reconnaissance version of the bomber, which were used for intelligence-gathering missions against Eastern Europe and the USSR. On 7 November 1958 a SAC RB-47H of the 55th SRS was flying over the Baltic when it was fired upon by Soviet fighters, but was not hit. On 1 July 1960 an RB-47H of the 98th SRW flew from Brize Norton on

Below: The USAF B-47 jet bombers first arrived at Brize Norton in September 1953, and continued regular deployments from the USA until 1958. B-47E 20581 of the 321st BW was at Brize Norton in early 1955. *Michael Bowyer*

an electronic reconnaissance sortie over the Barents Sea with the secret nuclear test site on Novaya Zemlya of particular interest. The aircraft was intercepted by a Soviet MiG-19 and shot down, only two of the crew of six surviving to be taken prisoner by the Russians and after a prison term, were sent home. In late 1959 the 308th BW deployed eleven B-47s at Brize Norton, supported by KC-97Gs. Various detachments continued into early 1960 with support from KC-135s and in January 1964, a B-58 Hustler, the world's first supersonic bomber, arrived from Carswell AFB in Texas. A small number of Hustlers followed, with reducing numbers of B-47s and B-52s, resulting in Brize Norton being handed back to the RAF on 1 April 1965 with the departure of the last B-47E of the 380th BW on 3 April 1965.

Entering service with the RAF were some new large aircraft, such as the VC-10s of 10 Squadron and Belfasts of 53 Squadron. The original strategic transport base had been at Lyneham with the Comet jet airliner fleet, but Brize Norton was selected as the new RAF base for global transport and freight. New passenger and cargo terminals were built, together with a 1,039.5ft long by 193.5ft wide cantilever hangar erected on the NE corner of the airfield with eleven bays capable of holding VC-10s and Belfasts when requiring maintenance. The new hangar was completed in June 1967 at a cost of nearly £2 million, and was the largest cantilever structure in Europe. The Belfasts and VC-10s of 53 and 10 Squadron moved in from Lyneham and Fairford in mid-1967.

The passenger VC-10s were supplemented by higher-capacity TriStars with 216 Squadron, and a batch of ex-BA Super VC-10s were modified at Filton as flight refuelling tankers for 101 Squadron, which were later joined by the converted 10 Squadron VC-10s. Six of the ex-BA TriStars were also capable of flight refuelling and carrying freight using a forward cargo door. Three similar TriStar 500s previously operated by PanAm were converted by Marshall at Cambridge to a purely passenger configuration. While the TriStars are expected to continue in service for some time, the aging VC-10 tanker fleet is due to be replaced by specially configured Airbus A330s under a £7.8bn Future Strategic Tanker Aircraft (FSTA) private initiative, with construction of a major new servicing facility at Brize Norton and the first A.330 airframe being modified by Airbus in Spain. Meanwhile after a long gap, the Belfasts were replaced by C-17 Globemaster IIIs operated by 99 Squadron in the strategic transport role, and with the planned closure of Lyneham, the C-130 Hercules fleet was expected to be based at Brize Norton until the types premature withdrawal announced in October 2010. The station will be the home of the new Airbus A400M strategic transport due to start replacing the older Hercules from around 2015. Brize Norton is therefore set to become probably one of the busiest RAF stations with round-the-clock operations of tankers, and passenger and cargo aircraft operating in support of RAF operations overseas.

Bruntingthorpe, Leicestershire;
SP599665, 10 miles S of Leicester.

Bruntingthorpe opened in November 1942 as a typical three-runway short-term wartime bomber station and was used as a Wellington-equipped bomber OCU as a satellite to North Luffenham. Aircraft accommodation was four T.2 hangars and one B.1, which still exists, with basic camp facilities for administration and living accommodation. No.29 OTU disbanded on 27 May 1945, and from December 1944 until 1947 the airfield was used for test flying of aircraft from nearby Bitteswell, the airfield then being allocated to C&M. The base was allocated to the USAF on 13 February 1957 and the main runway was extended to 10,500 feet for the operation of fully laden B-47s, the airfield supplementing the other four main SAC-allocated bases. In January 1959 the first B-47Es of the 100th BW arrived as part of operation Reflex Action, which was then at its peak, with small groups of Stratojets operated in the USA and overseas bases, although these deployments ceased in April 1959 at Bruntingthorpe. In September 1959, Bruntingthorpe became a satellite of 10 TRW at Alconbury with the 19th TRS RB-66Bs arriving, and staying until September 1962 after which the base was deactivated and returned to the RAF on 28 September and taken over by the motor industry for testing and storage. While still used by the motor industry, it is privately owned and the home of the British Aviation Heritage Collection, looking after a number of Cold War jet aircraft maintained in working order, and was also the base for the restoration of airworthy Vulcan B.2 XH558, although this is often maintained and operated from more suitable airfields.

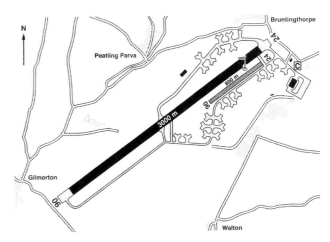

Above: Bruntingthorpe is now an unlicensed airfield which is heavily used for motor vehicle proving. The Cold War Jets collection is open most Sundays from 10.00 to 16.00 hrs and from time to time there is an open day when many of the aircraft are brought to life and taxied along the runway. *Pooley's Flight Guides*

Chelveston, Northants;
TL015685, 4 miles NE of Rushden

Construction of a wartime short-term bomber base commenced in 1940, with three runways connected by a concrete perimeter track giving access to 50 hardstandings. Aircraft could be serviced in one 'J' Type hangar supplemented by a pair of T.2s all located on the western boundary, between the brick-built flying-control Type 518/40. The majority of the camp and technical site consisted of rapid-build temporary brick buildings in addition to some prefabricated concrete structures. The Station HQ, Operations Block, stores and workshops within the Technical site were all of temporary brick construction, with additional accommodation in metal corrugated Nissen huts. On 15 August 1941 it was taken over by the RAF as a satellite to Polebrook until 2 September 1941, when it gained its independence and formed its own Station HQ. Work then started on the extension of the main runway to 2,000 yards which was completed in March 1942, the other two runways remaining at 1,400 yards.

The airfield was then allocated to the USAAF, but before they moved in, the new runway was used for glider trials by the Airborne Forces Experimental Establishment with Horsas being towed by Stirlings. On 10 June the Americans began to arrive with 10th and 11th Squadrons as part of the 60th Troop Carrier Group, with 48 C-47 Skytrains arriving between 20 and 28 July. The 60th TCG departed for Aldermaston on 7 August 1942 to make way for the B-17Fs of the 301st BG 8th USAAF from 9 August, the third squadron arriving on 16 August, and more B-17Fs on 26 August bringing the squadrons up to full strength. Meanwhile glider trials continued and some of the 352nd Squadron B-17s were dispersed to Podington, returning to Chelveston on 2 September to allow operations to commence on 5 September. The four squadrons continued in the daylight bombing campaign across Germany and occupied Europe before the 301st BG moved to North Africa. On 6 December 1942 the 305th BG began to transfer from Grafton Underwood while runway repairs were made. Leader of this group was Colonel Curtis LeMay, later to become a famous Air Force General. Immediately after arrival, the four squadrons of B-17s went into combat on the daylight bombing campaign, leading to the first attack on Germany on 27 January 1943, the target being the dockyard at Wilhelmshaven. The bombing campaign continued with the 305th providing support in the run-up to D-Day by attacking the V-1 launch sites, and after D-Day hitting tactical targets in support of the ground troops, followed by the 422nd Squadron flying H2S-equipped B-17s in a pathfinder role from 22 March until 23 May 1944. Combat operations ceased after the last raid on 25 April 1945 and the 305th left Chelveston for garrison duties in Belgium. The RAF took over the base in October as a satellite of 25 MU before being changed to C&M, with a small area allocated in 1951 to an RAF Radio unit.

During 1952 the base was allocated to the USAF and a new 9,000ft runway was built right through the middle of the original runway system for the operation of B-45 jet bombers, with an overrun taking it to almost 11,000 feet. This runway extended to the south of the original airfield, causing the closure of the Chelveston to Yeldon road. On 1 December 1952 the USAF took control to prepare the base for operations, and on 1 November 1955 Chelveston was transferred to the SAC as a stand-by base. From late 1958 to mid 1959 the 301st SRW RB-47s rotated through Chelveston and in September 1959 the base was returned to the 3rd AF. On 25 August 1959 the 10th Tactical Reconnaissance Wing arrived with RB-66Cs of the 42nd TRS, the aircraft being equipped with electronic jammers. Also the 47th BW brought in 13 B-66Bs which carried ECM systems in the bomb-bays, which were stand-off jammers to protect fighters. The 42nd TRS moved to France in March 1962 and in August Chelveston was reduced to inactive status before return to the RAF on 19 June 1964. From 1 January 1967, the base was used by the 10th TRW at Alconbury for storage of War Reserve Material in the 'J' Type hangar until the late 1980s, and the MOD closed the site in the early 1970s, the concrete runways, taxiways and hardstandings being broken up for aggregate. The 'J' Type hangar was the last building to be demolished and in late 2005, RAF Chelveston was sold by the Defence Estates Agency for development as a Renewable Energy Park and agricultural use. There is a small domestic estate near Caldecott, which is used by families of USAF personnel based at Molesworth. Although the former right of way across the runway extension has been restored, it is now as a bridleway.

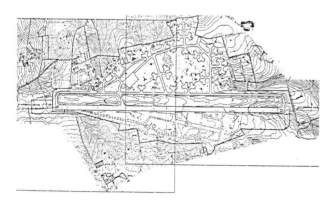

Above: Originally constructed as a short term RAF bomber station with three runways, Chelveston was completed by March 1942 and then handed over to the 8th USAAF from June 1942 until April 1945 when the station closed. It was returned to the USAF in 1952 and a new 9,000ft long main runway was constructed across the old airfield as a standby base. The first B-47s arrived in late 1958 and the main aircraft accommodation was a 'J' Type hangar on the western side with a very basic technical and accommodation site adjacent. In August 1962 flying ceased and the buildings and concrete were all removed. *RAF Museum*

Fairford, Gloucestershire;
SP150990. 2 miles S of Fairford town.

Now well known as the venue for the annual Royal International Air Tattoo (RIAT), Fairford has had a varied history with the RAF, USAF and as a Concorde flight test establishment. The original plan in 1943 was to construct a USAAF transport/air observation base, but instead it was earmarked for Bomber Command as an Operational Training Unit (OTU). However, it was too far south of the action, and on 18 January 1944 Fairford became part of 38 Group Transport Command with glider-towing Stirlings of 620 and 190 Squadrons, followed by Horsa gliders, to undertake cross-country exercises. Just before midnight on 5 June 1944, the first of 45 Stirlings carrying 887 paratroops of the 6th Airborne Division began to depart for Normandy where the task was to secure the Orne Bridge and the area east of the river, followed later on 6 June in the main Horsa glider landings near Ranville with the 6th Air Landing Brigade.

Fairford was constructed with three concrete runways and 52 spectacle-type hardstandings dispersed around a concrete perimeter track and two T.2 hangars were erected for essential maintenance. As the construction started late in World War 2, the buildings were mainly of a temporary nature, the technical site being in the NW side, and the bomb store on the opposite SE side. The station squadron, flight offices and operations block were of temporary brick construction and many other departments were accommodated in Nissen huts. The Stirling tugs occupied the dispersals and the disposable Horsa gliders were picketed normally without control surfaces to avoid storm damage.

The next major action for Fairford was on 17 September 1944 with the station's two squadrons of 38 Stirlings being pathfinders, while towing Horsas for the start of Operation Market Garden, the Arnhem operation. Another 43 Horsas were delivered on the second day, but on the third day the defences began to take their toll on the air operation. The worst day was 21 September, when Luftwaffe fighters and anti aircraft guns shot down seven of 190 Squadron's ten Stirlings and two more from 620 Squadron were also shot down. In October 16 crews from each squadron towed 32 Horsas across France to Rome, and by 10 October 27 had landed safely before moving on to Pomigliano near Naples. All the Stirlings returned to Britain in November, but to Great Dunmow, where the Fairford station organisation had also moved. Fairford became a Keevil satellite with 22 Heavy Glider Conversion Unit tasked to replace the pilots lost at Arnhem. The tugs were 29 Albemarles with a mixture of Horsas and the inferior Hadrian gliders, and training continued until 21 October 1945 when Fairford was placed on C&M. A number of training flights were operated from Fairford during 1945/46, but in September 1946 Fairford regained its Airborne forces role with four squadrons of 40 Halifaxes on experimental glider operations until December 1948, when the station returned to C&M.

In June 1950 Fairford was allocated to the Americans with the first personnel arriving on 7 July. Work then commenced on major developments which included lengthening, widening and strengthening runways and aircraft aprons with the construction of a special storage area for nuclear weapons. The erection of three Luria hangars began in June 1951 and on 1 July, the base was officially transferred to American control, in particular the 7th Air Division SAC. On 7 February 1953 17 of the enormous and noisy B-36 bombers arrived from Carswell AFB, Texas, two of the aircraft failing to arrive due to accidents.

The B-47B Stratojet made its operational debut in Britain when two arrived at Fairford on 7 April 1953 for a short stay in advance of the first B-47 Wing of 45 Stratojets of the 30th BW which arrived on 2 June 1953, the three squadrons returning to the USA on 4 September. The arrival of the B-47s started

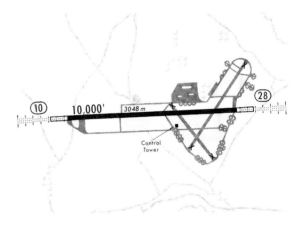

Above: Construction of Fairford commenced in 1943 for the USAAF, but it was taken over by the RAF for troop carrying glider operations in January 1944. Three runways were built with 52 dispersals connected by a perimeter track and a very basic technical site with two T.2 hangars on the north side of the airfield. It was reduced to C&M in December 1948, and allocated to the USAF in June 1950 when considerable improvements were made for the operation of B-36s which first arrived in February 1953. The main runway was extended to 10,000 feet and the others closed to flying operations. Three Laria hangars had been erected in mid-1951 to complement the two T.2s, and B-47s began to operate from Fairford from April 1953, until the station was returned to RAF control in June 1964. It then became the base for the Red Arrows Gnats, as well as Belfast, VC-10s and initial RAF Hercules operations until they departed for Lyneham in February 1971. The long runway and unobstructed approaches were ideal for Concorde flight testing from April 1969 until January 1977, and runway repairs were made from April 1977 for the operation of USAF KC-135s from September 1978 until January 1990. Fairford was then put on a reserve status, the old concrete runways and taxiways were breaking up and the whole base was closed for major improvements between 2000 and 2002. It remains the venue for the annual RIAT. The March 1964 approach chart shows the main runway extension to 10,000 feet with the technical site on the north side and a modern air traffic control tower on the south side of the runway. *Jeppesen*

Left: From the end of the Cold War, Fairford has been used by USAF B-52s on deployments from the USA covering trouble spots in Europe and the Middle East. B-52H 60-0052 of 11 BS/2 BW was a participant in the RIAT flying programme at Fairford in July 2006. *Author*

the replacement of the SAC piston-engine bomber fleet, and with a top speed of over 600 mph they were faster than many of the jet fighters in service with either side. At the beginning of 1953 there were only 62 B-47s in SAC service, but by the end of the year there were 329, which was slightly more than the B-36 fleet. The three-crew B-47's pilot and co-pilot were located in tandem under a canopy on the long slim fuselage with a navigator in the nose, who handled weapons release. The co-pilot was kept busy handling the electrical systems and flight engineering duties, as well as remotely controlling a pair of radar-directed 20mm cannons in the tail. It had thin wings swept back at 35 degrees with power from four podded GE J-47 turbojets giving a high subsonic cruising speed. With a weight fully loaded of ninety tons including a weapons load of 20,000lb, take-off could be boosted by 18 solid fuel rocket-assisted take-off (RATO) bottles, mounted nine on each side of the fuselage, which gave an additional 18,000lb of thrust. The Stratojet had a novel undercarriage arrangement with two pairs of main wheels in tandem under the fuselage and outriggers under each wing. The taller nose wheel was also used for steering on the ground and cross wind take-offs were challenging. The aircraft literally flew off at 140 knots, as the take-off attitude was the same as it sat on the ground with no back pressure on the control column. The B-47 had an operational ceiling of over 40,000 feet and on internal fuel the range was 3,000 miles, which could be extended by flight refuelling, including taking fuel from a KB-47B Stratojet, which was available well in advance of the KC-135s based on the Boeing 707 jet airliner. As well as its take-off, the B-47 had a demanding landing technique, which began with a long straight-in approach from high altitude, and when the undercarriage was lowered it acted like an air-brake losing height at a rate of 5,000 per minute. Flaps were lowered on the final approach, which began several miles from the runway threshold and the approach required

One of the USAF hangars at Fairford with Waddington Wing Vulcan B.2 XH557 visiting for the IAT in July 1973. *Author*

great concentration, as the speed had to be kept within two knots of 130 knots without stalling, to prevent an over-run of the runway, as each knot above the landing speed added 500 feet to the landing run. The ideal was to touch down on both main wheels together, because if only one made contact, the aircraft would bounce back into the air. To slow the fast-moving bomber, a brake parachute was deployed immediately on touchdown to complement the normal heavy braking, which had an anti-skid device. The average B-47 landing roll was along about 7,000 feet of runway, which is why the SAC runways were at least 10,000 feet long.

B-47 deployments continued, including some RB-47Es of the 68th SRW and a few RB-36Fs of the 5th RW visited Fairford in July 1954. Also in the summer of 1954 three squadrons of B-47E from the 43rd BW stayed for three months, but the SAC use of Fairford reduced from the autumn of 1955. In January 1958 the 21-day Reflex deployments started with small batches of B-47s staying at one base for three weeks before moving elsewhere, to prevent heavy losses in the event of a pre-emptive strike. With the end of the Reflex deployments, Fairford reverted to RAF control on 26 June 1964.

The Central Flying School (CFS) at Little Rissington used Fairford as a relief landing ground, the Gnat Trainers arriving at the beginning of 1965, staying until September 1966. The RAF aerobatic team, the Red Arrows formed on 1 March 1965, and was based at Fairford until 1 August 1965. At the other extreme in size were the Belfasts of 53 Squadron which re-formed at Fairford on 1 November 1965 before moving into the custom-built facilities at Brize Norton in May 1967. During work on the runway at Brize Norton, the VC-10s of 10 Squadron also made use of Fairford. After a quiet summer Air Support Command took over Fairford in September for the introduction of the C-130K Hercules to the RAF. On 25 February 1968 47 Squadron re-formed, followed by 30 Squadron in June with both squadrons operating trooping and cargo flights. The last Hercules left for Lyneham on 10 February 1971 and Fairford reverted to being a RLG and satellite for Brize Norton.

Meanwhile due to the runway at Filton being inadequate for the full flight development testing of Concorde, on 9 April 1969, Fairford was designated as the Flight Test Centre of the supersonic airliner with its 10,023ft runway 27/09 with unobstructed approaches at both ends. Concorde 002 made its maiden flight from Filton on 9 April 1969 and landed at Fairford at the end of the initial test, to be followed later by the British-built pre-production aircraft and production examples until the lease expired on 31 January 1977.

There was a need for runway and taxiway resurfacing and repairs which commenced in April 1977, with Fairford classed as a reserve airfield in September. Then came the announcement in July 1978 that Fairford was to become the second British location for USAF KC-135 tankers, and on 12 September the first five arrived for an autumn exercise, and after flying up to four sorties daily, they departed back to the USA on 28 September. On 1 February 1979 preparations were started to make Fairford capable of taking six-week rotations of KC-135s and the larger KC-10s, together with short deployments of B-52s. On 1 June 1979, the base was designated a USAFE Prime Installation and on 13 September the first five KC-135 deployment arrived, followed by small numbers over the next decade and by March 1981 the full strength of 15 tankers was reached, the most vital task being the aerial refuelling of boom receptacle-equipped USAFE and NATO aircraft. Although not part of the Cold War, Fairford became busy on 14/15 April 1986 with three KC-10As from 2nd BW and KC-135s drawn from eight wings to support the F-111s attacking Gaddafi's Libyan military installations. In January 1990 all USAFE tanker operations were transferred to Mildenhall, and Fairford became a forward base in 1991 for B-52s operating over Iraq in Desert Storm, ten B-52s operating from Fairford from 1 February until the last sortie on 25 February, the wing deactivating on 6 March. With the concrete runways and taxiways crumbling there was an urgent need for construction work, and the base received a major rebuild of the airfield facilities between 2000 and 2002. New dispersals were constructed and special hangars to house the B.2s, one of the few global dispersed sites for this advanced stealth bomber. Fairford has now been reduced to a reserve status, after having a major investment in its facilities, but is expected to continue for the foreseeable future as the home of the RIAT.

Greenham Common, Berks;
SU500645, 2 miles SE of Newbury by A339

Greenham Common is probably best known for its Cold War cruise missiles and Women's Peace Camp, the latter remaining long after the nuclear missiles had gone and the site returned to peaceful activities. The ladies had obviously abandoned their previous homes and the entrance to Greenham Common was where they lived in caravans.

The site of the airfield was requisitioned in May 1941, and work soon started on creating a bomber OTU as a satellite of Aldermaston. The main east-west runway 6,000 feet/1830m long was built along the ridge with shorter secondary runways of 4,200 feet/1,280m and 3,300 feet/1,006m, restricted by the topography. The main technical site was to the south of the airfield and the bomb store and sewage works were located on the NE corner. The runways were linked by a concrete perimeter track with 50 multi-engine aircraft hardstandings and two T.2 hangars were erected for aircraft maintenance. Accommodation was widely dispersed, particularly the base housing. Buildings were to the basic wartime standards with temporary brick and pre-cast concrete construction, supplemented by Nissen huts. By the time the station was ready for occupation, priorities had changed and it was allocated to

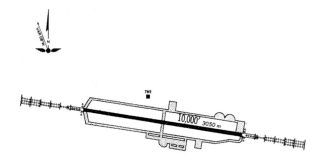

Above: The March 1964 approach chart shows the 10,000ft main runway constructed along the ridge for SAC jet bomber operations. The main technical site was on the south side of the airfield and later cruise missile bunkers were built at the western end. With operations ceasing in March 1991, the base closed and has been largely returned to common land with the runway and perimeter tracks removed. *Jeppesen*

the USAAF, although Greenham was used for RAF flying training until the end of September 1943, the USAAF officially taking over on 1 October.

The first flying unit was the 354th FG of the IXth AF which arrived on 4 November to convert from Airacobras to the P-51B Merlin-powered Mustangs, as the first unit to fly the type in combat. However, its stay was short as it moved to Boxted on 13 November. The IXth AF was newly created to be the spearhead of the invasion of Europe and a number of fighter groups stayed for short periods before it was decided to allocate the base to tactical transports, and the C-47s of the 438th TCG arrived on 16 March 1944 as part of the 53rd TCW. Since arriving in Britain the TCW had been confused as they had been resident at Grantham and then Langar with their aircraft eventually reaching them at Langar in mid-February, but all their supplies were at Barkston Heath, and then they were moved again to Greenham. Despite this confusion the unit was declared operational in April 1944, fully trained in paratroop drops and glider towing. Large areas of steel mesh were laid at both ends of the main runway to assist with the marshalling of the gliders, and it was at this time that the extended-length T.2 hangars were erected to store and service the glider tow cables.

The 438th was then allocated the task of being the spearhead of the airborne forces on D-Day, the 5/6 June. The first C-47 was airborne at 22.48 hrs on 5 June, carrying 1,500 paratroops of the 101st Airborne Division, followed by the towing of Hadrian and Horsa troop-carrying gliders, earning the 438th a Distinguished Unit Citation. As the land forces advanced across Normandy, as with other TCGs, the missions changed to resupply drops, and as rudimentary landing strips became available, the C-47s delivered cargo and returned with casualties. Ninety aircraft were used to drop the 101st around Eindhoven in Operation Market Garden on 17 September, and the following day a further 80 aircraft towed gliders to the same area without loss. The TCG supported the allies in the Battle of the Bulge, and in the Ardennes in mid-December 1944. When the 438th moved to France in February 1945, Greenham Common was left with no flying units and was closed on 1 June 1946.

Although there were plans to return the site to common land, the warming up of the Cold War brought a requirement for USAF nuclear bomber bases and despite local objections, work started on the construction of the longest runway in Europe on 21 April 1951. All the wartime buildings, including the T.2 hangars, were demolished and a large new technical site built with the A339 diverted around the site. The new 11,000ft/3,048m runway was built on the site of the old main runway extended in both directions with parallel taxiways, and by September 1953 the new base was ready to accept the SAC B-47s. The first deployment was in March 1954, when the first aircraft of the 303rd BW arrived from Davis Monthan AFB, although after the first aircraft had landed, the runway was found to have deteriorated and the remaining B-47s diverted to Fairford. The cause was found to be a re-routed stream which had reverted to its original course, and the trapped B-47 was flown out of the shortened runway with the benefit of rocket JATO bottles and a light load of fuel.

Further substantial runway reconstruction was made, together with additional hardstandings for up to 57 aircraft, with improved drainage to the site. The work continued until October 1956 when 45 B-47s of the 310st BW began SAC's full ninety-day deployment to Greenham Common. The local residents were objecting to the high level of noise of the B-47s both with ground engine running and on approach and departure. Opposition increased considerably when a B-47 developed a problem on take-off and dropped the underwing fuel tanks, which landed on a hangar and parked B-47. Two USAF personnel died with two more injured, and the fire was not extinguished for 16 hours. When the 90-day deployments were changed to three-week Reflex Alert rotations the noise reduced, but further concerns were voiced when more work was done to strengthen the runway and dispersals to accommodate the B-52 Stratofortresses, which apart from occasional visits were never based there. A rare visitor on 16 October 1958 was the supersonic B-58 Hustler, which had flown non-stop the 8,028 miles/12,929km from Tokyo to Greenham with five in-flight refuellings in a time of 8 hr, 35 min and 20 sec.

With the end of Reflex rotations on 1 April 1964, Greenham Common was returned to the RAF on 1 July, but the forced withdrawal of the American military from France by General de Gaulle renewed interest in the base and it was further upgraded as a NATO standby base. The next residents were the 20th TFW F-111E from Upper Heyford, arriving in March 1976 while the runways were being resurfaced at their home base. In 1973 what became the Royal International Air Tattoo moved from North Weald to Greenham Common, then known as the Embassy Air Tattoo after its sponsors. The Tattoo, bringing the largest collection of military aircraft from

Left: The base was completely rebuilt after WW2 for SAC jet bomber deployments and the hangars had central slots to give clearance for aircraft tails. One of the visitors for the IAT at Fairford in July 1973 was Wisconsin ANG KC-97 Stratotanker 0-20905. Underwing podded jet engines helped to increase the speed before the arrival of the KC-135s, and the aircraft is fitted with the boom flight refuelling system used by the USAF. *Author*

around the globe, was an annual event until 1977, when it was run every two years to alternate with the Farnborough Air Show, and due to airfield developments the Tattoo moved to its current home at Fairford.

The USAF wanted to use Greenham Common as a base for KC-135 tankers, but following strong public opposition this plan was abandoned even though further work was carried out on the runway and buildings. However, the decision was made by NATO in December 1979 to use the base for nuclear-armed Ground Launched Cruise Missiles (GLCM), the other one being Molesworth. Six protective shelters were built on the SW corner of the base to house ninety-six missiles and their launchers, to be deployed widely around the countryside in the event of the threat of aggression. With the major advances in micro-electronics and miniaturisation in the 1970s new smart weapons systems were developed, with the cruise missile in the forefront. The cruise missiles were designed to fly at low heights above the terrain and at subsonic speed, making them a difficult target for radars to detect, especially as they never flew in a straight line for long periods. The cruise missile follows a pre-planned route to its target, the directing computers using an advanced TERCOM system (Terrain Contour Matching). The information required by this system is provided by reconnaissance s defining the contours and topographical features that the missile will pass on its way to the target, which are pre-programmed into the flight computer. The computer is linked to a TV camera located in the missile nose, allowing comparison with the pre-programmed images and allowing course corrections to be made. The images for the TERCOM system took many years of satellite reconnaissance flights over potential target routes by the American Big Bird and KH-11 Key Hole spy satellites. Extremely high-resolution images were obtained with details as small as 30cm being clearly visible which meant that small targets such as buildings could easily be destroyed by a cruise missile.

Although the noise complaints ceased, the Campaign for Nuclear Disarmament (CND) objected and the Women's Peace Camp was set up at the main entrance to the base in October 1981. While work continued on the construction of the nuclear installations, the protest grew, culminating in some 30,000 women linking arms around the nine-mile base perimeter on 12 December 1982. Despite the obstructions and publicity, C-5 Galaxy freighters began to arrive in May 1983 to deliver the equipment for the 501st Tactical Missile Wing (TMW), with the missiles beginning to arrive from November. The new technical site, known as GAMA (GLCM Alert & Maintenance Area), had two Launch Control Centres (LCC), and four Transporter Erector Launchers (TEL) in each of the low-profile shelters. The protests grew even stronger when the 501st began to practise operational deployments to secret pre-set potential launch sites. At the peak of the protests there were some 40 women's peace camps around the perimeter of the base.

When the USA and Soviet Union signed the Intermediate-range Nuclear Forces Treaty in 1987, led to the removal from Greenham Common of the GLCMs in March 1991, the 501st TMW being deactivated on 4 June, and the airfield was handed back to the RAF. In February 1993 the site was put up for sale, although the Peace Camp remained until September 2000. The former USAF buildings have become the basis for a business park and large areas of the common have been returned to their natural condition, including the removal of most of the runway and hardstandings for use as hardcore in the Newbury by-pass. The control tower and the missile bunkers remain derelict and unused as a reminder of the final Cold War use of the site.

Lakenheath, Suffolk;
TL740820, 4 miles SW of Brandon, by A1065

Lakenheath is now one of the most modern USAFE strike fighter bases and occupied by the 48th TFW equipped with two squadrons of F-15Cs and one squadron of F-15E Strike Eagles. In 1941 work started on the preparation of a satellite to Mildenhall to replace Newmarket, the site coming into use prior to completion with the Wellingtons of Honington-

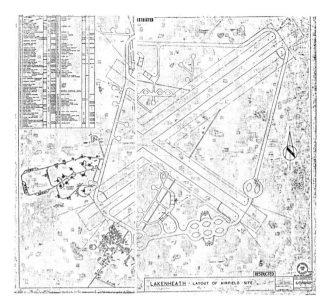

Left: Lakenheath opened in November 1941 as a satellite to Mildenhall for operations as a bomber OTU. It had the traditional three-runway layout with a very basic technical site on the southwest side of the station, since the main support was at nearby Mildenhall. A weapons store was located on the western boundary and hangarage consisted of a B.1 and two T.2s located on the northern side of the airfield. The station was taken over by the USAF in August 1948 and the main runway extended to 9,000 feet for operations by B-29s and the massive B-36s, followed by B-47s. Soon after the plan of the airfield was produced in 1952, the main runway was further extended resulting in a diversion of the Brandon road. The 48th TFW arrived from France with F-100D Super Sabres, to be replaced progressively by F-4Ds, F-111Fs and currently F-15Cs and F-15E Strike Eagle. *RAF Museum*

based IX Squadron. Due to the poor state of the landing area at Lossiemouth, the Wellingtons of 20 OTU used Lakenheath from 24 November 1941 until 12 January 1942. In January 1942 149 Squadron with Stirlings began to fly from Lakenheath, training in the operation of the type and then flying operations including mine laying, and followed with the supply flights for the resistance movement in occupied Europe.

The airfield was constructed to the usual bomber pattern with three concrete runways linked by a perimeter track, the main runway being 6,000ft long. There were 36 heavy bomber hardstandings around the perimeter. Hangarage consisted of two T.2s and one 'B1' in the technical site located at the south end of the airfield. Administrative accommodation was very basic in a collection of huts, as the main base was at nearby Mildenhall. The facilities began to have some improvement in 1942 and in April 1943 Horsa gliders began to arrive for storage at the southern side of the station. On 20 June 199 Squadron arrived with Wellingtons which were replaced by Stirlings from 5 July, continuing in the heavy bombing role until 1 December 1943 when they were taken off operations to train in the use of a radar jamming system called Mandrel. In May 1944 both 199 and 149 Squadrons moved out of Lakenheath.

With the departure of the Stirlings, Lakenheath was selected as a very heavy bomber base, but the preparations took so long that the war came to an end before the work was finished. The airfield was used for various systems trials without any based aircraft, but that changed on 5 August 1948 when Lakenheath became active again with the USAF taking over residence, which has continued ever since. Three squadrons of 32 B-29s of the 2nd BG arrived and over the next eight years some 30

5.43 The 48TFW used 2-seat F-100Fs for conversion and continuation training, 3856 being seen in natural metal finish on the approach to Lakenheath in July 1967. *Author*

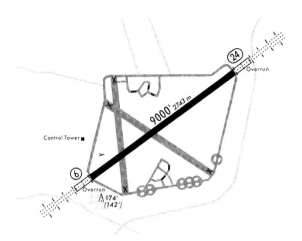

Above: Construction of Lakenheath commenced in 1941 as a satellite to Mildenhall to the RAF bomber pattern with three concrete runways, the main one being 6,000 feet long, and 36 heavy bomber dispersals around the perimeter track. The rudimentary technical site with two T.2 and one B.1 hangars was located at the southern end of the airfield. The RAF bombers departed in May 1944 to allow major developments for very heavy bombers, but continued until after the end of World War 2. The USAF arrived in August 1948 with B-29s and B-50s, the main runway having been extended in 1945 to 9,000 feet and the other two runways extended to 2,000 feet, requiring the Brandon road to be diverted around the site in 1952. With the American withdrawal from France in October 1959, the base was transferred from SAC to USAFE control and the F-100Ds of the 48 TFW arrived in January 1960. These were followed by F-4D Phantoms, F-111Fs, during which time HASs were erected, and the current F-15Cs and F-15E Strike Eagles which arrived from February 1992. The May 1968 approach chart shows the 9,000 feet main runway, with the other two disused and the diverted Brandon Road. *Jeppesen*

Bomb Groups and Bomb Wings rotated through Lakenheath on 90-day deployments. For these operations all three runways were extended in 1945, the main runway reaching 9,000 feet and the other two 6,000 feet. In addition to the technical site, a large domestic site was added to the immediate south of the base with extensive married quarters, and the usual PX and Commissary, together with entertainment facilities which for a time included a very well-equipped casino, as well as medical facilities. As the base expanded a further domestic site was constructed on the old RAF training station at Feltwell to the north of Lakenheath.

The B-29s of the 2nd BG were replaced by the 22nd BG on 19 November 1948 and the B-50A was introduced to Lakenheath on 20 August when the 65th BS of the 43rd BG arrived, supported by three KB-29M flight refuelling tankers. Supporting transports included C-47s, C-54s, C-97s and the large-for-its-time C-74 Globemaster with its side-by-side twin cockpit canopies. With the heightened tension of the Korean crisis, the B-29s doubled their tour and to achieve greater load-carrying capability and range they were lightened by the removal of armament and non-essential equipment. In mid-1950 Lakenheath was declared the best US base in Britain with development potential and was earmarked as a fighter base until Manston was selected as an alternative. However, new dispersal pads were approved for up to twelve B-36s and five hangars were modified as nose docks, the first of which arrived briefly from the 7th and 11th BGs on 15 January 1951. In the same month B-50Ds of the 93rd BG also arrived, supported by the boom-equipped KB-29P tankers, together with RB-50Bs of the 55th SRW to produce radar maps of Eastern Europe. Also visiting were some of the first C-124 Globemaster IIs to arrive in Britain. Three more of the enormous and noisy B-36Ds arrived in June 1951 to make a formation flypast at the Paris Air Show on 1 July. The B-45As of the 91st SRW were the first jet bombers at Lakenheath between September and October 1951.

By 1952 there had been further developments at the base, with the Brandon road diverted around to the east and because of the nuclear weapons stored in the north-west of the base, security was improved by replacing the barbed wire with a ten-feet-high wire fence surrounding the perimeter. The base was still being used by reconnaissance RB-36s and RB-50s when the first B-47s visited on 13 April 1953, and in August 1953 16 F-84G jet fighters of the 508th SFW practised rapid deployments. On 1 January the base closed for further improvements, no doubt in preparation for the arrival of the highly classified Lockheed U-2, two of which arrived in May 1956 from the 1st Weather Reconnaissance Squadron (Provisional), but attracted so much local aviation enthusiast interest that they left for Wiesbaden in late June. In 1954 KC-97 tankers arrived for the first time and in June the B-47 deployments started, which continued until 1956 when the strategic bombers were pulled back to the Midlands. Both Mildenhall and Lakenheath were used from 15 March 1958 for the delivery of Thor IRBMs for the RAF using C-124s and C-133s.

With the withdrawal from France by US forces, Lakenheath was transferred from SAC to USAFE on 1 October 1959 as part of the operation Red Richard, the USAF departure from France. Development work was undertaken at the base, including the removal of the hump in the main runway and the provision of protected shelters to allow the nuclear-armed F-100Ds to start the engine for a rapid scramble in the event of an approaching enemy attack. On 5 January 1960 the 48th TFW began their move from France to Lakenheath, where they have now been in residence for over half a century. They brought with them 75 F-100D Super Sabres which operated from Lakenheath until 1972, when they were replaced by F-4D Phantom IIs. The first F-4D arrived on 7 January 1972, and the last F-100D departed on 15 April by which time only 17 Phantoms had arrived, which was well below the required strength due to delays in arrival of the new type. It was not until September 1974 that the 494th TFS was fully operational, followed by 492nd TFS on 10 December and the 493rd TFS on 13 January 1975.

The stay of the Phantoms was brought to a premature halt by the announcement that the 48th TFW was to re-equip with F-111Fs of the 366th TFW, the conversion period being from 14 October 1976 to 31 December 1977. This change was a result of Soviet deployment of SS-20 IRBMs (Intermediate Range Ballistic Missiles), the first being set up in the spring of 1977. The American government required a F-111E/F nuclear strike force based in Europe in support of NATO. All the F-111s were fitted with Pave Tack to allow the launch of small tactical missiles round the clock in all weathers with laser-guided precision to the targets. The first three F-111Fs arrived on 1 March 1977 and the last F-4D departed on 22 April, after which the Wing had stood down to prepare for the operation of the new swing-wing strike aircraft. On 1 April the 495th TFS reactivated with the first of up to 84 F-111Fs, the three designated roles being nuclear strike, defence suppression and naval support. Each aircraft was capable of carrying up to 14,000lb of external war load, including two nuclear weapons, and in June 1977 the first main delivery of a total complement of 91 F-111Fs was made with completion the following month. The 494th TFS was operational on 9 July followed by the 493rd on 1 August, the 495th concentrating on the training task and the 492nd not becoming operational until 31 August 1978. The F-111s were equipped with APQ-119 advanced navigation and attack radar as well as a very accurate low-level terrain-following system, allowing almost unhindered penetration of hostile airspace, until the advent of stealth bombers such as the Northrop B-2.

In the late 1970s further changes to the airfield included the erection of Hardened Aircraft Shelters (HAS) as well as new concrete aircraft platforms and it was not until the early 1980s before all the based aircraft were in HAS. Many of the 48th TFW aircraft deployed overseas, while the home-based aircraft practised Tacevals, full scale alerts and hi-low profile training sorties around Scotland. The 1986 raid on Libya proved the high level of readiness of the F-111F crews, which was further confirmed with 18 F-111s departing for Taif in Saudi Arabia on 25 August 1990 to respond to the Iraq invasion of Kuwait. A total of 66 F-111s were deployed for strikes in Desert Storm in January 1991 and by the end the 48th had flown about 2,200 sorties dropping conventional 1,000lb and 2,000lb precision-guided Laser Guided Bombs (LGB) against a variety of targets, all the F-111s returning without loss to Lakenheath by mid-May 1991.

With the far superior Soviet threat from their new generation strike aircraft such as the MiG-25 Foxbat and Sukhoi fighters armed with radar guided air-to-air missiles based not just in the USSR, but also East Germany, the USAF did not have an aircraft which could beat the Soviet threat, until the development of the phenomenal F-15A Eagle. Powered by two P&W F-100 turbofans, with full fuel and armed with four AIM-7F Sparrow fire and forget air-to-air missiles, the thrust to weight ratio was better than 1:1. This was demonstrated in February 1975 when an F-15A broke eight climb-to-height records, during one of which, the aircraft exceeded the speed of sound 19 seconds after take-off and continuing to climb vertically, passed through Mach 2 in two minutes. The improved F-15C could be armed with either four AIM-7Fs or AIM-120A AMRAAMs (Advanced Medium Range Air-to-Air Missiles), plus four AIM-9M Sidewinder IR heat seeking missiles and a GE six-barrel 900 round 20 mm rotary cannon. The F-15Cs are equipped with Hughes APG-63 radar which can detect hostile aircraft up to 100 miles away and illuminate them for the fire control computer, which directs the missiles against low and high level targets.

Although the Cold War was over, the 48th TFW was not allowed to relax and on 21 February 1992 the first F-15E Strike Eagle arrived to start the replacement of the F-111s with the last to depart being on 18 December 1992. The 492nd and 494th TFS were each issued with 24 F-15Es, and on 14 December the 493rd TFS was deactivated until late 1993 when a batch of F-15Cs were delivered for a reactivated 493rd TFS. Meanwhile in early 1993 the first two squadrons were declared operational and undertook deployments to Incirlik in Turkey protecting the Kurdish minority in Iraq. The F-15s continue to operate from Lakenheath, which is the major USAFE strike base outside the USA and is co-located, but independent from nearby Mildenhall.

Right: The replacement of the 48th TFW F-111Fs started with the arrival of the F-15E Strike Eagles in February 1992, equipping both the 492nd and 494th TFS. USAFE F-15E 90-0248 carrying the tail markings of the 48th TFW deploys its large airbrake on landing at Fairford in July 1995. The aircraft is fitted with conformal fuel tanks alongside the fuselage under the wing to increase the combat radius. *Author*

Manston, Kent;
TR333662, 3 miles NW of Ramsgate on A253.

The airfield of Manston first came into use in early 1916 when the initial temporary hangar and wooden huts were erected in February to provide a defence of London against bombing by Zeppelin airships and later Gothas. In the spring of 1917 a training school was set up to prepare pilots for combat flying

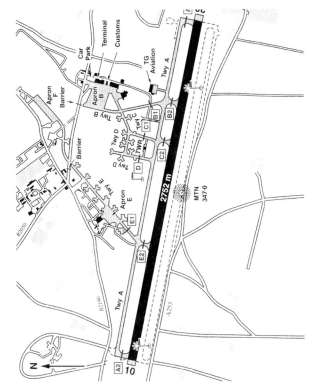

Above: The airfield at Manston was first used in 1916 for the air defence of London from German bombers and airships. It became a permanent station in 1917 and was used for flying and technical training until the start of World War 2. Fighter Command took over in November 1939 and the station was badly damaged by enemy action during the Battle of Britain. Due to the growing number of damaged aircraft needing a safe haven to land, work commenced on a new 9,000ft/2,739m runway, which was also 750ft/229m wide and had a 6,000ft/1,829m loop at the western end to drag the wrecked aircraft off the runway, allowing more arrivals. It was ready for operations from April 1944 and 616 Squadron with the RAF's first jet fighters, Meteor F.1s arrived in July 1944 to help destroy the V-1 flying bombs. The airfield was transferred to the USAF in July 1950 as an advanced SAC fighter escort base with F-84s arriving from November 1951. These were replaced by F-86s including radar equipped F-86Ds from November 1953, until the station was handed back to the RAF in May 1958. With no regular RAF flying units, commercial flying started in April 1959 and has continued with varying success to the present with the runway reduced to half its original width. The RAF station finally closed in March 1999, but a MOD presence remains with the Fire Service Central Training Establishment. *Pooley's Flight Guides*

in France, while the War Flight patrolled the area between Ramsgate and Whitstable attempting to provide day and night protection. With the decision to make Manston a permanent station in 1917, there was a major expansion of hutted accommodation and hangars. With the formation of the RAF on 1 April 1918 a bomber-training unit arrived and the War School supplied pilots for combat in France. By the end of World War 1, there were eight large and one small aeroplane sheds, two MT sheds and three large workshops with many wooden huts covering 100 acres out of the total of 680 acres of the site. The observers' school continued with a reduced capacity until closure in September 1919 and flying ceased on 7 February 1920, but a School of Technical Training (SoTT) was opened with courses commencing in May 1920. Regular flying returned with the arrival of 6 FTS to provide refresher courses for pilots being posted overseas, but the FTS was closed in April 1922, and the SoTT courses were reduced. The station hosted a variety of fighter and bomber units and with the expansion of the RAF in the mid 1930s the enlarged SoTT became No.3 SoTT and Manston became a ground and flying training establishment.

With the declaration of war, the training units moved out, the deserted station being put temporarily under C&M until 235 and 253 Squadrons were formed initially equipped with Battles, both squadrons moving out in February. Manston was transferred to Fighter Command on 15 November 1939 and the first resident fighter unit was 79 Squadron with Hurricanes which arrived on 12 November and 600 Squadron with Blenheim night fighters before the end of the year. During and after the Battle of France a number of squadrons passed through with 600 Squadron as the main resident. The station was right in the front line during the Battle of Britain as an

Above: The USAF was based at Manston from November 1951 until it was returned to the RAF control in May 1958. Being a very old RAF station mainly designed for technical training and badly damaged during the Battle of Britain, Manston was not built to the traditional RAF Expansion Scheme layout and the various dispersals were constructed for the USAF fighters. The large emergency runway runs along below the border of the photo. *RAF Museum P.017588*

Left: The 406th FBW operated F-84Es at Manston from July 1952 until November 1953, when it was redesignated the 406th FIW with F-86 Sabres. The flight line includes 92049 which was present in September 1952, soon after the unit arrived. *Michael Bowyer*

advanced airfield in the Hornchurch Sector of 11 Group with 54 and 74 Squadrons detached daily from the sector station. Being closer to the front line than any other station, Manston attracted the destructive attention of the Luftwaffe during the Battle of Britain with the first light bombing on 3 July 1940. However on 12 August there were concentrated attacks on Manston as well as Lympne and Hawkinge. Just as 65 Squadron were about to take off, Bf.110s began to shoot up the airfield from low level with Do.17s bombing from a higher altitude. All but one of the Spitfires were able to take off while a total of some 150 bombs were dropped, destroying workshops, damaging two hangars and cratering the landing ground. One civilian was killed, one Blenheim destroyed and a number of Spitfires damaged. The next attack came on 14 August when Bf.110s dive-bombed the airfield, destroying three Blenheims and badly damaging four hangars with the landing area further cratered. Enemy attacks continued on the station throughout the month, leading to a major attack on 24 August when the first wave of bombers caused such a large cloud of dust and chalk that the second wave bombed Ramsgate with a devastating loss of life and property. The third wave of Ju.88s killed seven with many more injured, many buildings wrecked, and the water main broken, leaving fire crews short of water to fight the blazes. All external contact was lost for several hours with the telephone lines breached, and while the station personnel were trying to clear up the mess, a further raid took place. As a result Manston was abandoned apart from being an Emergency Landing Ground (ELG) for the rest of the Battle. With the Battle of Britain over, Manston was reoccupied by cannon-armed Spitfires of 92 Squadron in January 1941 and with its advanced strategic location took the offensive against targets in occupied Europe. With both fighters and light bombers, the station was returned to a full operational status in October 1941, although hard runways were still to be constructed. Being so close to the Continent, the station was a haven for aircraft in distress, from fighters to heavy bombers, many of which crash-landed on the airfield.

Finally the need for an emergency runway was acknowledged and work started on 15 June 1943 to the south of the existing landing ground on a 9,000ft/2,734m by 750 feet/229m concrete runway with a 6,000ft/1,829m dispersal loop at the western end containing twelve crash bays where wreckage could be dragged to maintain a clear runway. Ground-attack rocket- and cannon-armed Hurricanes and Typhoons continued to operate from the northern section of the airfield with Manston becoming the top-scoring station within 11 Group. The massive new runway was ready for use on 5 April 1944 and in the first three weeks 56 emergency landings were made. With the approach of D-Day on 6 June, Manston became even more active, with emergency arrivals continuing and the station being in the front line of the V.1 flying bomb offensive.

On 20 July 1944 Manston had the arrival of the first two Meteors for 616 Squadron, the only Allied jet fighter to take part in World War 2. As more Meteors became available, the squadron relinquished their Spitfires and were declared operational on 26 July. The squadron claimed its first V.1 a week later, but as the guns had jammed, the pilot toppled the V.1

Below: The 406th FIW operated F-86D Sabre night fighters at Manston from the end of 1954 until May 1958 when Javelins with the RAF were able to take responsibility for the night air defence of Britain. F-86D 24092 of the 514th FIS was one of the aircraft-based at Manston. *Michael Bowyer*

by lifting his wing under the wing of the bomb, the disturbed airflow causing the Doodlebug to topple and crash. Tempests of 501 Squadron concentrated on anti-diver night operations and claimed for 33 by the end of August, the advanced location of Manston allowing the RAF fighters to attack the flying bombs over the Channel in advance of the coastal anti-aircraft guns. With the overrunning of the V.1 launch sites by the advancing allies in Europe, the station turned once again to the offensive with up to eight squadrons operating Tempests, Meteors, Mosquitos, Spitfires, Beaufighters and Avengers.

The new runway also had the FIDO fog-dispersal system installed alongside which consisted of burning petrol, the heat of which improved visibility for landing aircraft in fog. It was first really used in anger on 21 September 1944 when over a six-hour period nine Spitfires, two Mustangs, three Norsemen and a Stirling were able to make emergency landings, with many more taking advantage throughout the winter months. The last offensive operation was on 25 April by Spitfires providing escort for Lancasters and Halifax bombers. After the war Manston became somewhat quieter and was taken over by Transport Command as a staging post for aircraft flying to the Middle East.

However on 11 July 1950 the station was transferred from the RAF to the 3rd Air Division USAF to be used as a base for a SAC fighter-escort wing. Manston was once again a UK frontline fighter base when the F-84Es of the 20th FBW arrived as the first of a series of deployments. From November 1951 when the USAF took over control, new hardstandings were constructed, and the width of the runway was enough to allow it to be divided into three strips with the centre 200ft/61m wide section used for landing and take-offs and the strips either side for taxiing aircraft. In November 1951 the 123rd FBW of the Kentucky Air National Guard (ANG) became the first ANG unit to be deployed to Britain with F-84Es to replace the 12th Fighter Escort Wing, and ANG personnel were rotated through Manston until July 1952 when the 123rd was deactivated and its aircraft used to form the 406th FBW. In November 1953 the 406th FIW re-equipped with F-86 Sabres, replacing them with the all-weather radar-equipped F-86D at the end of 1954. The F-86D radar was developed to carry out radar-controlled interceptions with an automatic fire control system which automatically computed a 90-degree collision course with the target. When the target was within range up to 24 Mighty Mouse rockets were fired in a salvo, each rocket having the equivalent destructive force of a 75mm artillery shell, the rocket armament replacing the 20mm cannons. The F-86Ds were provided as an interim measure by the USAF for the night air defence of Britain until the arrival in service of the more advanced Javelin all-weather fighters with the RAF. The 406th FIW was therefore deactivated in May 1958 and the station returned to RAF control as a Master Diversion airfield from April 1956 open 24 hours a day, 365 days a year to accept aircraft in an emergency. Manston was reduced to C&M on 1 August, the airfield closing until reopened again as a Master Diversion airfield on 28 March 1959, but with no residential flying units. From April 1959 Manston began to be used by commercial flights with the main RAF presence being the RAF Fire Training School using a number of redundant airframes. With a declining requirement, RAF Manston closed on 31 March 1999, although the MOD retained the Fire Service Central Training Establishment, and the airfield became Manston International Airport with mixed success. The Battle of Britain connections are remembered by the Spitfire and Hurricane Memorial Building, housing a preserved example of each aircraft.

Mildenhall, Suffolk;
TL685768, 12 miles NW of Bury St Edmunds by A1101.

Having only briefly served during the early Cold War with RAF Lincoln bombers, the main Cold War and after operations have been with the USAFE. On 15 July 1950 the B-50Ds of the 329th BS, 93rd BG arrived with increased security as some of the aircraft carried unarmed nuclear weapons due to the heightened tension of the Korean War. The B-50Ds, while looking similar to the earlier B-29s, was a 75 percent new design with a top speed of 400 mph at 25,000 feet and could reach 40,000 feet with an effective bomb load. Its range with

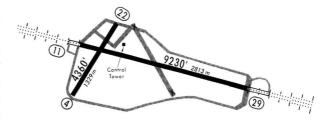

Above: Mildenhall first opened in October 1934 after construction started in 1933 on a pre-RAF Expansion Scheme Bomber station. The technical site was located on the northwest corner and included two 'A' Type and three 'C' Type hangars, the technical, administrative and accommodation buildings being of brick construction. The airfield closed in November 1942 for the construction of three runways with 36 heavy bomber dispersals, and reopened in April 1943 for RAF heavy bomber operations which continued after the end of World War 2 with Lincoln B.2s from July 1949 until March 1950. The station was then closed for the lengthening of the main runway in 1951 to 9,230 feet and other improvements made ready for the arrival of the Americans in July 1950 in preparation for B-50 operations. In 1959 Mildenhall became the main logistics gateway to Britain and a new passenger terminal was built. The November 1967 chart shows the cross runway 22-04 still available, but this was closed and used for the dispersal of the SR-71 Blackbirds from September 1971 until they were withdrawn at the end of 1989. Mildenhall continues as a major USAF logistics base, and the 100th ARW with KC-135R tankers has been resident since January 1992. *Jeppesen*

Right: The first of the high-speed SR-71A Blackbird high-altitude reconnaissance aircraft, 17972 of the 9th SRW, arrived at Mildenhall on 9 September 1974, having flown from New York to Farnborough in a record time of 1 hour, 54 minutes and 56.4 seconds. *Author*

a 10,000lb bomb load, equivalent to a single first-generation nuclear bomb was 6,000 miles, although it could lift up to 28,000lbs of bombs over shorter ranges.

Early in 1951 the main runway was further extended and in February the B-29s of the 509th BG arrived for a three-month deployment, the rotation of which included KC-97E tankers for the first time, tankers continuing in operation to the present day. From late 1957 B-47s were based at Mildenhall and until August 1959 B-47s with their take-offs boosted by JATO bottles strapped to the side of the fuselage were based there to be able to carry heavy weapons loads. At the beginning of 1959 Mildenhall was selected as the main USAF gateway to the UK, replacing Burtonwood, which was located too far from other USAFE bases and suffered from poor weather and runway subsidence. Mildenhall continues as the major USAFE transport terminal in the UK and the first arrival of many was a C-124C Globemaster II in January 1959.

With the base being underutilised, Mildenhall became the European C-130 Hercules base from 1 June 1965 with the units operating two-month deployments. The strategic transports continued to visit, the first by a C-141 being 19 October 1965, and although there was not a direct combat role for the base, its Cold War intelligence-gathering duties started with the arrival of the first EC-135 in November 1965, and at the end of 1969, the ELINT RC-135s moved to Mildenhall from Upper Heyford. The RC-135U/V/Ws have numerous blade antennas under the fuselage and large fairings on the forward side of the fuselage housing advanced Side Looking Airborne Radar (SLAR) which can receive and analyse radio and radar signals from the Soviet Bloc and Eastern Europe, allowing the development of appropriate countermeasures. On 16 January 1970 Mildenhall became the UK base for European KC-135 tanker operations, and the first of many C-5A Galaxies visited on 20 December 1970. Following its record flight to Farnborough on 9 September 1971 of the futuristic SR-71 Blackbird, the aircraft became the

Mildenhall also became the European base for the specially-equipped intelligence-gathering C-135s. USAF EC-135H 61-0291 of 10th ACCS, 513rd ACCW fitted with a flight refuelling boom makes a smoky take-off from Mildenhall in May 1991. *Author*

Above: Probably part of the 90-day rotation programme at Mildenhall was C-130E AF64-560 FB visiting Alconbury on 21 October 1972. *Author*

first of a number of its type to be based at Mildenhall, together with some U-2s from the spring of 1976. These aircraft were Detachment 4 of the 9th SRW and were tasked to carry out strategic reconnaissance missions for NATO and the US Department of Defence. In addition to the regular reconnaissance flights, the aircraft were able to gather telemetry signals from Soviet weapon systems, assisting the US Defence Intelligence Agency to obtain good information on the Soviet advances in nuclear weapons. The Blackbird flights were generally along the borders of Eastern Europe, with additional flights to the Russian naval facilities around Murmansk and the Kola Peninsula. SR-71s remained in service until the end of 1989 when the US DoD withdrew them for budgetary reasons, but there are rumours that a number have been stored in operational condition at a secret base in the Nevada Desert.

Right: Molesworth originally opened in May 1941 as a RAF bomber station with three runways and a single B.1 hangar on the northerly located technical site, the station was allocated to the 8th USAAF in May 1942 following which the main runway was extended and two T.2 hangar added ready for operations to commence in June 1942. The B-17s arrived in September 1942 and continued operations until April 1945, when the airfield was handed back to the RAF until it was placed on C&M in October 1946. The airfield was again selected by the USAF and the main runway extended at the eastern end as shown on the 1958 plan, although the base had been deactivated in December 1957. It became designated as the second UK cruise missile base in 1988, but was never active before the end of the Cold War. It is still a USAFE base and apart from the cruise missile structures, much of the remainder of the base has been cleared, including the runways. *RAF Museum*

The C-130 rotations were gradually reduced by January 1976, leaving Mildenhall with the tasks of supporting a flight refuelling tanker fleet and the operation of a transport terminal for which new facilities were built to the east of the existing technical site. By the start of the 1980s a SR-71A of Detachment 4 of the 9th SRW was based in a secure area on the south side of the main runway, and in April 1984 the British Government announced approval for two SR-71s to be based at Mildenhall. To provide support with the special fuel required by these Mach 3 80,000 feet altitude titanium aircraft, up to eight KC-135Q tankers were based at Mildenhall until the withdrawal in January 1990. Also seen at Mildenhall during this period were a small number of TR-1s, a development of the U-2, also used for intelligence gathering flights. The base continues to be a major military passenger and cargo terminal and the resident KC-135Rs are operated by the 100th ARW which was activated on 13 January 1992.

Molesworth, Cambs;
TL008775, N of A14, 11 miles NW of Huntingdon

The hilltop site of Molesworth with excellent approaches was planned from 1939 as an RAF bomber base for the Liberators of 8 Group with three runways and a basic technical site located to the north of the flying field with a single 'J' Type

hangar. The station was opened on 15 May 1941, but it was 15 November before 460 (RAAF) Squadron formed with Wellington IV bombers, but no operations were flown before the squadron departed for Breighton on 5 January 1942. Little further happened at Molesworth until a visit by General Ira Eaker with his staff to consider the suitability of the base for the operation of B-17s. As a result there were major improvements together with an extension of the main runway, improved accommodation and the erection of two T.2 hangars, and the advanced party of Americans arrived on 12 May. The first regular residents was the 15th BS and having no aircraft of its own, was attached to 226 Squadron RAF and borrowed some Boston IIIs, ready for the first USAAF bombing raid on a European target on 29 June 1942, hitting Hazebrouck marshalling yards. The 15th BS acquired their own ex RAF Bostons in July and marked them with the US star and bar. With plans for Molesworth to become a B-17 base, and the first was noted in August and the 15th BS left on 13 September, while the 303rd BG began to move in on 12 September. After working up, the 303rd flew their first operation on 17 November with 16 B-17Fs against U-boat pens at St Nazaire. The 303rd, known as "Hell's Angels" continued on 8th AF daylight raids, including their first raid on Germany on 27 January 1943, when they bombed Wilhelmshaven. One of the B-17Fs of 358th BS, named "Hell's Angels" was the first B-17 to complete 25 missions over Europe, although the "Memphis Belle" was awarded the honours as the film team were located at Bassingbourn and the two aircraft were neck and neck for the achievement. Another B-17 known as "Knockout Dropper" was the first to achieve 75 sorties with the 359th BS without any injuries to the crews. The 303rd flew its final combat sortie on 25 April 1945 and in early June left for North Africa where it was allocated to transport duties.

Molesworth was returned to the RAF on 1 July 1945 and on 16 July 441 and 442 (RCAF) Squadrons arrived with Mustangs, and on 27 July Meteor IIIs of 1335 CU moved in from Colerne to convert piston fighter pilots to jet aircraft. Both the Canadian squadrons disbanded on 10 August, and were replaced by the Mustangs of 234 and 19 Squadrons followed by some Spitfire units, before the arrival on 6 October 1945 of the Meteor IIIs of 124 Squadron, which stayed until 6 October. In late October 222 Squadron arrived from Weston Zoyland for conversion to Meteors and left for Exeter on 11 December, to be followed by 234 Squadron in mid-February, which left for Boxted in March when the conversion was completed. After 54 Squadron operated Tempests from Molesworth in September and October 1946, the station was placed under C&M.

With its excellent approaches, Molesworth was again selected by the Americans in July 1951 and a new long runway was constructed across the previous three runway layout with other improvements ready for the arrival of 582nd Air Resupply Group with a dozen B-29As together with three RATO-equipped C-119Cs and an assortment of other aircraft. On 25 October 1956 the 582nd ARG became the new 42nd Troop Carrier Squadron equipped with just transport aircraft, the B-29s having departed. On 31 May 1957 the TCS was placed under the control of Alconbury and deactivated on 8 December 1957. Having the status as a reserve airfield, as well as a supply depot for the USAF, transport aircraft continued to use the airfield and some WB-47s operated from the airfield in the late 1950s before it was closed to flying.

In the 1980s Molesworth became famous as the second base for the ground-launched cruise missiles with a compound containing four shelters each with a capability of launching 16 missiles from pre-planned sites in the surrounding countryside. The construction of the missile facilities across the western end of the main runway was completed in 1988, but these were decommissioned soon after with the end of the Cold War. There was a much smaller Peace Camp by one of the entrances, the remains of which still exist, and Molesworth is still used by the USAFE for intelligence-gathering from images taken by reconnaissance aircraft operating over Europe, with the domestic site at Alconbury.

Sculthorpe, Norfolk;
TF860315, 5 miles W of Fakenham by A148/B1454

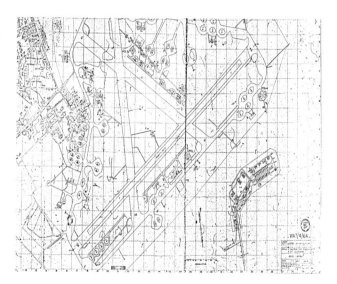

Above: First opened in June 1943 as an RAF bomber station, Sculthorpe featured the usual three-runway layout, but with very basic facilities including two T.2 and one B.1 hangars in the technical site on the western side of the airfield. From October 1943 the station was used by 140 Wing Mosquito FB.VIs on low-level precision raids across occupied Europe, but was put under C&M after World War 2 until the USAF arrived in January 1949. The main runway was extended to 8,950 feet for the operation of B-29s and B-50s, followed by B-45 reconnaissance jet bombers which continued until June 1962, when the station was put on a stand-by basis. The 1965 plan of the site was after active operations had ceased. *RAF Museum*

Above: USAF C-119 51-2619 was with the 49th ABG at Sculthorpe in 1954. *Michael Bowyer*

The remote airfield at Sculthorpe was completed as a typical utility mid-war RAF bomber station with a small basic technical and accommodation site on the western boundary and three concrete runways with an interlinking perimeter track giving access to 36 heavy bomber dispersals. Two T.2 and one B1 hangar were erected and the first aircraft to arrive were 32 Horsa gliders in January for storage in the open. The airfield became fully active when 342 (Free French) Squadron arrived in June 1943 with Boston IIIs and Havocs for training, after which they departed in July. On 20 and 21 July 464 (RAAF) and 487 (RNZAF) Squadrons arrived at Sculthorpe where they were soon training on Mosquito FB.VIs to be used on precision raids in support of the invasion, especially at night. They were joined by 21 Squadron on 27 September. Their first operation was in October, led by AVM Basil Embry, to attack power stations in France, and in case of capture Embry was disguised as Wg Cdr Smith. The three Mosquito squadrons were combined under 140 Airfield with a capability of rapid deployment when required, and in December the destruction of the V.1 launch sites began, with the squadrons moving into Hunsdon on 31 December 1943 to be closer to the action. Being sited in the area of 100 Group, Sculthorpe transferred from 2 Group and on 17 January 1944 214 Squadron arrived, equipping gradually with ex-USAAF B-17s commencing operations on the night of 20/21 April dropping bombs on French railway marshalling yards. No 214 Squadron moved out in May and Sculthorpe closed for conversion to a very heavy bomber base, which put it out of action for the rest of the war.

Immediately after reopening in December 1948 it was put under C&M and in October it had been offered to the Americans, initially allocated to SAC bombers, and on 17 January 1949 the first members of the USAF arrived to begin preparations, the main runway being 8,950 feet long. The first arrivals were the B-29s of the 92nd BG from 7 February 1949, staying until 18 May, when they were replaced by three squadrons of B-29s of the 98th BG until 15 August. On 22 August the first B-50As arrived of the 63rd Squadron,

Left: TB-26 41-39278 was based at Sculthorpe in 1954 with 5th TTS on target towing duties. *Michael Bowyer*

43rd BG supported by three KB-29M tankers. Rotations of B-29s and B-50s continued, including ten RB-29s of the 23rd Strategic Reconnaissance Squadron of the 5th SRG which operated intelligence-gathering flights around the Iron Curtain boundaries until February 1950, later to be replaced by the RB-29s of the 72nd SRS.

The intelligence gathering was further developed from May 1951 when a detachment of the 91st SRG brought a batch of RB-45C Tornado reconnaissance jets to Sculthorpe which remained until 1955. An unusual and at the time secret operation was two flights by RAF-crewed RB-45Cs with British markings by three aircraft in over ten-hour sorties over the Soviet Union at night. The object was to observe ICBM sites and other strategic targets with 35mm cameras taking images of the radar scopes identifying ground features. The first operation was in April 1952 and it was repeated again two years later. The SAC deployments continued until November 1951 when the 22nd BG departed, the base becoming home for a short while to a rescue squadron which departed to Prestwick.

The first permanent basing of USAFE aircraft at Sculthorpe were the B-45As of the 84th and 85th BS of the 47th BG (Light) which began to arrive on 5 June 1952 flying across the North Atlantic via Greenland and Iceland. On 20 December 1953 the 422nd Squadron joined the 47th BG and in March was renumbered the 86th BS, moving to Alconbury on 20 September 1955. The Tornados were each capable of carrying up to five 2,000kg tactical nuclear weapons, and the Tornado compliment was increased on 11 May 1954 when the RB-45Cs of the 19th TRS of the 66th TRW arrived, remaining until 10 January 1959 when it moved to Germany. The 47th BG had up to 75 B-45s based at Sculthorpe with a nuclear strike role and in January 1958 RB-66s began to replace the RB-45Cs with the 19th Squadron, soon followed by the 47th BS. The twin-jet B-66 could carry a bomb load of 12,000lb over a combat radius of 800 miles on internal fuel and had a maximum speed of 620 mph at 10,000 feet. In support of the Wing, boom-equipped KB-29P tankers had arrived in January 1956, the earlier KB-29s using the British-developed hose-and-drogue method of transferring the fuel in the air. In January 1957 KB-50Ds began to replace the earlier tankers, and in April 1958 they were replaced by the faster KB-50Js with jet engines in pods under the wings. The last seven of these aircraft returned to America on 22 March 1964. On 22 June 1962 the 47th BW had already been deactivated, ending the offensive role of the base and putting it on a standby status. Following the end of the Cold War, it became a storage site for redundant French Air Force Super Mysteres and F-100s which had been purchased by USA Mutual Defence Aid funds. In April 1967 Sculthorpe became a Standby Dispersal Base for the 48th TFW and the main runway was resurfaced in 1968. It remains in occasional use by military aircraft while part of the technical site is now a business and leisure park.

Shepherd's Grove, Suffolk;
TM990730, 9 miles NE of Bury St Edmunds by A143.

One of the last World War 2 combat airfields to be completed, Shepherd's Grove was built in 1943 as a temporary bomber base for USAF use. However it opened instead on 3 April 1944

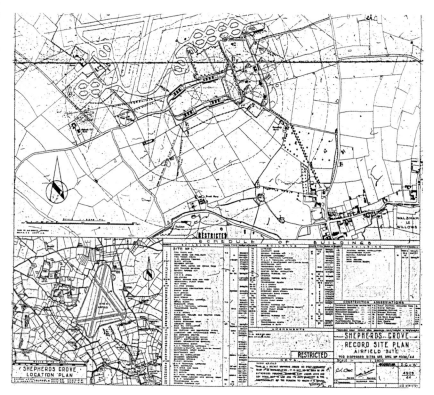

Right: Little changed at Shepherd's Grove from its 1944 configuration until the arrival of the 81st FIW F-86 Sabres in August 1951. The airfield opened in April 1943 as a temporary base for the USAAF with three runways and a rudimentary technical and domestic site on the western boundary of the airfield, including one T.2 hangar, with a second T.2 in the northeast corner. The F-86 Sabres flew from the airfield until December 1958 when flying ceased and it became a Thor missile base from April 1959 until the missiles were withdrawn in June 1963. The reclad T.2 hangar on the technical site still exists, as does the northern end of the main runway with associated perimeter track. *RAF Museum*

Above: Shepherd's Grove was the base where the USAF F-86A Sabres of the 81st FIW arrived on 27 August 1951 from their transatlantic crossing, as the first swept-wing fighters to be stationed in Britain. The aircraft were allocated to the air defence of Britain under the control of RAF Fighter Command and a pair of the F-86As visited de Havilland at Hatfield in April 1952.
de Havilland

under the control of 3 Group RAF as a satellite to Stradishall. From 14 May to 5 October 1944 Stirlings of 1657 Conversion Unit were operated from the airfield. Three concrete runways were constructed with the rudimentary technical site on the western boundary including two T.2 hangars, one of which still exists with some Nissen huts and other World War 2 buildings. The runways were connected by a concrete perimeter track with 50 bomber-type hardstandings. In late 1944 Shepherd's Grove was transferred to 38 Group, Transport Command, still with Stirlings of 196 and 299 Squadrons which arrived at the end of January 1945 in support of SOE supply drops and to train for future airborne operations. On 24 March 1945, the Stirlings were used to tow Horsa gliders from Shepherd's Grove as their contribution to the crossing of the Rhine. Both squadrons remained after the war carrying passengers, mail and cargo on overseas flights until 1946 when 299 Squadron disbanded on 15 February, followed by 196 Squadron on 16 March. There was little operating from the airfield until flying ceased in February 1950 and the station was placed on C&M.

However, there was a requirement for the USAF to have permanent escort fighter bases in the UK to protect the B-29 operations. With the start of the Soviet-supported Korean War in July 1950, tension increased further, prompting plans to move up to four fighter escort wings to Britain, the proposed F-86 Sabre jet requiring at least a 6,000ft runway. In August 1950 Britain offered airfields at Bentwaters, Wethersfield and Shepherd's Grove, with interim bases at Bassingbourn, Lakenheath and Sculthorpe also available. In November the decision was made to make Manston the main fighter base with Bentwaters, Wethersfield and Shepherd's Grove as interim airfields. Shepherd's Grove came under the control of 11 Group Fighter Command on 12 February 1951 and work commenced on 16 March with preparation by the Americans for the arrival of F-86 squadrons. The F-86 Sabre was a fast and highly manoeuvrable air-superiority jet fighter which had been a great success in the Korean War. In the period from 1950 to 1953 USAF F-86 Sabre pilots claimed at least 792 North Korean fighters which were mostly MiG-15s for the loss of 78 Sabres.

With a well publicised arrival on 27 August 1951, the 116th Fighter Interception (FIS) Air National Guard (ANG), 81st FIW arrived overhead in a mass formation after a ferry flight across the North Atlantic in their F-86As, the first swept-wing jet fighters to be based in Britain. On 5 September they were joined by the 92nd FIS, while a third squadron flew into Bentwaters, and the 116 FIS was renumbered the 78th FIS on 1 November 1952. These were the first foreign units to be allocated to the air defence of Britain under the control of Fighter Command since the end of World War 2. On 1 March 1954 the 81st FIW became part of the 49th Air Division with HQ at Sculthorpe and all three squadrons changed their role on 1 April when the 81st FIW became the 81st Fighter Bomber Wing (FBW) with air defence as a secondary task. For the new role the F-86s were replaced by F-84F Thunderstreaks with the 81st non-operational from 22 April until 8 February 1955, while training on the new type took place. These aircraft were capable of carrying nuclear bombs and take-off could be boosted by rocket-assisted take-off (RATO) gear. With the last Sabres departing on 8 February 1955, the re-designation as FBS was completed. The 92nd FBS moved to Manston on 28 March 1955, leaving the 25 F-84Fs of the 78th FBS in residence at Shepherd's Grove. Between 9 August 1955 and 27 April 1956 the F-100s of the 77th FBS deployed to Shepherd's Grove while their runway at Wethersfield was resurfaced, and in turn the 78th FBS lodged at Sculthorpe from 31 May 1956 to 3 May 1957, while the runway was also resurfaced. On 1 July 1956, with the 49th AD disbanding, the 81st was designated a Tactical Fighter Wing (TFW) as part of the 3rd Air Force, USAFE. The 78th TFS moved out of Shepherd's Grove on 22 December 1958 to allow the construction of jointly British- and USA-operated Thor IRBM nuclear missile sites as a satellite to Feltwell, the RAF taking control on 1 April 1959.

On 22 July 1959 "B" Flight of 77 Squadron was allocated to Shepherd's Grove and as the missiles began to arrive, it was renumbered 82 Squadron on 1 September 1959, each squadron being equipped with three missiles. Any launch order, whether for training or combat, had to be authenticated using a top secret code by joint RAF and USAF teams at High Wycombe, where both RAF Bomber Command and USAF 7th Air Division were located. The preparation of Thor for launch required at least 105 minutes, resulting in a considerable warning required of impending enemy attack while the missile was raised to the launch position, aligned with the target position, and the missile was fuelled with liquid propellants. Additional time would also be needed to fit the nuclear warheads in times of tension, which were stored at the World War 2 airfield of Faldingworth near Scampton. With the Thors deployed in the open, they were considered vulnerable to attack and on 30 June 1963 the missiles were withdrawn. Up to 1966 the base was used by a number of USAF support organisations. It is still used as a housing annex for Mildenhall and Lakenheath, with all trace of the airfield removed apart from the north end extension of the main runway.

Upper Heyford, Oxon;
SP515260 5 miles NW of Bicester.

The landing ground was first prepared in 1916 and on 20 January 1918 the Canadian-manned 123 Squadron arrived with Sopwith Dolphins, re-numbered 2 Squadron Canadian AF in 1919 before departing to Shoreham. A second, similarly-equipped 81 Squadron formed on 20 November 1918, and after being renumbered 1 Squadron Canadian AF also moved to Shoreham, and Upper Heyford closed in 1920.

Following the 1923 Defence Review, Upper Heyford was selected for development as a bomber station with a major building programme in 1925, and despite being incomplete, reopened on 12 October 1927. Upper Heyford was really the prototype of the layout in the RAF Expansion Scheme, which was then adapted throughout the 1930s. The station was designed to accommodate three single-engine day bomber squadrons, the perceived enemy at the time being France.

The basic technical site located on the SE side of the landing ground had six 'A' Type hangars erected with four in a convex semi-circle and the other two behind the outer ones on the front row. The station HQ was opposite the guard room at the entrance. Included in the main site were an operations block, main stores and workshops, motor transport section, power house and water tower, all regular features of permanent RAF stations. A watch office was located between the hangars with a fire station and R/T communications station with gun butts, engine test house, bulk fuel and lubricants stores also provided. Across the road from the airfield was an impressive domestic site with five Type 'C' barrack blocks built around a large parade ground. Close to the entrance were the sergeants' mess and institute, with the dining room and cook house, sick quarters, church, cinema and assembly hall all close by. To encourage

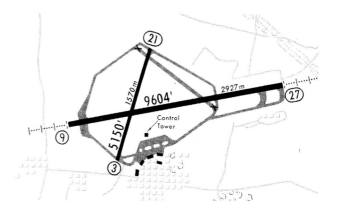

Above: After selection as an RAF bomber station in 1923, building started on the site of Upper Heyford in 1925 as a prototype of the RAF Expansion Scheme layout, ready for opening in October 1927. The technical site on the southeast boundary of the airfield included six 'A' Type hangars in a convex semi-circle, with four at the front and the other two behind the end units. The technical, administration buildings and the officers' mess were of brick construction behind the hangars, and the accommodation and recreation facilities were built on the south side of a local road outside the main camp. The first RAF bombers arrived in January 1928 and continued until the end of World War 2, during which time three concrete runways were constructed, the main runway being 6,000 feet long, with a perimeter track giving access to 23 bomber dispersals. Upper Heyford was allocated to the USAF in June 1950 as a strategic bomber base and amongst the major developments, the main runway was extended to 9,000 feet with the taxiways widened and strengthened. Operations commenced from December 1951, and further developments were in 1952, including another extension of the main runway to 9,604 feet with 21-03 still in use in August 1964, but later disused. A secure nuclear weapons maintenance and storage site was constructed on the north side of the site. The first B-47s arrived in June 1953, and remained on rotational deployments until 1967. In September 1970 the 20th TFW arrived with F-111Es and 50 HAS were erected dispersed around the airfield. The last F-111Es departed in December 1993 and the base closed in December 1993. Upper Heyford is now a designated Cold War Heritage Site by English Heritage, but entry by visitors is by prior appointment. *Jeppesen*

Above: Close to the trigger store was the nuclear weapons maintenance building. In addition to the watch tower, there were also tall poles erected around the site to stop landings of helicopters. *Author*

sporting activity, two football fields and recreational area were provided and the airmen's married quarters were located on one side of the domestic site, while the officers' mess and married quarters were located on the airfield site.

By the time the station had opened, the perceived French threat no longer existed and it became a two-squadron heavy bomber base with the first arrivals being Hyderabads of 99 Squadron on 5 January 1928. They were joined by 10 Squadron which re-formed on 3 January with its first three Hyderabads arriving on 28 January, starting a long period as a heavy bomber station. With the formation of 40 Squadron on 1 April 1930 with Fairey Gordons, 10 Squadron moved out to Boscombe Down, and 18 Squadron re-formed on 20 October 1931 with Hawker Harts, joined by the Harts of 57 Squadron on 5 September, with 40 Squadron moved to Abingdon in October 1932. On 14 November 1933, 99 Squadron began to replace the antiquated Harts with the first two Heyfords with the RAF and a year later moved to Mildenhall, making room for the Harts of 33 Squadron. A number of RAF bomber squadrons rotated through Upper Heyford, with 57 Squadron receiving Blenheims in March and April 1938, joined by 18 Squadron in May and 34 Squadron in July, the three squadrons being allocated to AASF in France during the Munich crisis. In January 1939 Upper Heyford became part of 2 Group Bomber Command and with 34 Squadron departed for Watton, the two Upper Heyford squadron-based Blenheims were allocated to the Air Component of the BEF to provide support for the British Army in France. With the outbreak of war, Upper Heyford became part of 6 Group and with both Blenheim squadrons posted to France, the station role was changed to bomber training with Wellingtons arriving from April 1942, and Mosquitos in December 1944, until the end of the war.

Upper Heyford was then allocated to the Americans as one of the four major USAF bomber bases, the first personnel arriving in June 1950 to begin preparations, the others being Brize Norton, Fairford and Greenham Common. The original airfield had been equipped with the usual three-concrete-runway layout with the main runway 6,000ft long and the perimeter track serving 23 bomber dispersals. The B-29s required a runway of 9,000 feet with over-runs, to allow for the later jet bombers in development which would require 10,000 feet. To accommodate the new American bombers the taxi tracks were widened to 100 feet and all the hard areas were strengthened to at least two feet thick. The four barrack blocks and airmens' mess provided initial accommodation for the American teams updating the base facilities including new domestic buildings, recreational facilities and the customary PX and commissary which continued almost to the final closure of the base.

The first operational unit was the 93rd Air Refuelling Squadron with KB-29Ps on 1 December 1951 on the first 90-day Temporary Duty (TDY) deployment. The return to the USA was usually during the last few days of the third month, the replacements arriving at the start of the following month. Further construction was undertaken in 1952 and included substantial strengthening of the airfield defences. Flight refuelling units continued to rotate through Upper Heyford and in June 1953 the first B-47B Stratojet bombers in Britain began to use the base for training, the initial resident B-47Bs being of the 22nd BW which were present from December 1953 to March 1954, and soon after the first B-47Es of the 303rd BW arrived. Although there were visits by the enormous reconnaissance RB-36F and H of the 5th SRW, the major residents of the base were the nuclear-armed B-47s including RB-47Es, which continued until 1967. The RB-36 contained a pressurised compartment with 14 cameras and a range of surveillance equipment and the crew compliment was increased from 16 to 22, the additional six members operating the reconnaissance systems. The long-range reconnaissance role was taken over by the RB-47Es from 1954, the Stratojet with seven cameras for day and night graphy and the RB-47H and RB-4Ks which had a pressurised compartment in the bomb-bay for electronic intelligence gathering operated by three specialists. In August 1962 three U-2s arrived at Upper Heyford, having flown the 3,000 miles from near New York in seven hours, causing concern regarding their planned activities. However, the explanation that they were to carry out a high-altitude sampling programme for nuclear debris was true, as the Soviets had just conducted a series of nuclear tests at the Novaya Zemlya site, including a mammoth 57-megaton hydrogen bomb. The U-2 sampling activities of the distribution, movement, retention and total amount of debris in the upper atmosphere allowing the US intelligence agencies to compile an accurate estimate of the yield and performance of the devises that had been tested.

On 8 February 1965 Upper Heyford became a Dispersed Operating Base (DOB), but that changed on 1 September 1966 with the arrival of RF-101 Voodoos of the 17th and 18th TRS, 66th TRW from France returning Upper Heyford to a Main Operating Base. In February 1969 the 17th TRS began to convert to RF-4C Phantoms and was transferred on 12 January 1970 to the 26 TRW at Zweibrucken as the 66th TRW was inactivated on 1 April. Voodoos were retained by the 18th TRS until 1 April, when they returned to the USA, and the base was prepared for its new equipment with the 20th TFW which moved from Wethersfield. Many improvements were made to the base and on 12 September 1970 the first two swing-wing F-111Es arrived, the 20th TFW reaching its full compliment by 29 July 1971 and was declared operational in November. The F-111E was a very advanced aircraft for its time, one of the unique features being a self contained zero-speed, zero-altitude emergency escape module. The two-man crew of pilot and weapon-system officer were accommodated side-by-side in an air-conditioned and pressurised cabin, and entry was via an upward-opening section of the canopy, hinged at the centre-line. In the event of an emergency both crew members remained in the cabin module, which is ejected away from the stricken aircraft by a 40,000lb-thrust rocket and lowered to the surface by a large parachute. Air bags cushion the impact on land, while providing flotation bags if the landing was on water. The emergency procedure could also be initiated from under water. With no ejection seats the crew have a high level of comfort without a complex strapping-in process, and all the systems and controls are easily accessible, with adequate room for manuals, charts and documents.

Although normally flown at very low levels, the performance of the F-111 was exceptional with a maximum

F-111E AF 68 022 of the Upper Heyford-based 20th TFW at the IAT Greenham Common in July 1976. *Author*

HASs in the QRA area where the F-111s and crews were kept at readiness. *Author*

speed at high altitude of Mach 2.5 with the wings swept to a maximum of 72.5 degrees. Using the maximum combined thrust of the two P&W TF-30 turbofans, the aircraft could climb at Mach 2. The crews of the 20 TFW were trained at Upper Heyford, the primary role being very low-level interdiction to remain below the enemy radars and guidance of the effective SAM missiles. The target may only be identified on a radar screen, with the weapon released and finding the target by inertial guidance avoiding the aircraft having to be too close. At the heart of the aircraft systems was a powerful computer which processes all the information from the aircraft systems and present it to the crew. The GE APQ-113 J-band multi-mode radar provided accurate air and ground navigation, ranging the weapons delivery to the target. In addition to the strike weapons, the F-111 was protected from hostile aircraft by air-to-air Sidewinder missiles. An essential system was the fully automatic APQ-110 terrain-following radar (TFR), which allowed the aircraft to follow the ground contours at a height of 200 feet at a speed of Mach 1.2, the radar sensing the topography ahead of the aircraft and making adjustments to ensure clearance of any obstacles. With everything automated, the crew's main task was monitoring the progress of the aircraft when flying in total darkness with no outside visual references. The main duty of the pilot is to monitor the TFR display, while the WSO monitors the ground-mapping radar to ensure that the correct route is being followed. In the stores management system, weapons were assigned to a particular target before take-off, and when the release point was reached, the relevant weapons were released automatically.

On 1 January a new unit was activated within the 20th TFW, the 42nd Electronic Countermeasures Squadron (ECS) equipped with the special EF-111A Raven ECM version of the bomber, the first of which arrived on 3 February 1984. Further changes occurred at Upper Heyford with the erection of 50 HAS to provide shelter for the F-111s in the event of a conventional attack. Security was increased with concrete blocks slowing traffic and anti-terrorist spiked rollers to deter ramming the gates. It was at that time that my wife and I were the guests of her US Marines son who took us into the commissary, but I had strict instructions not to speak as an English accent would not have been welcome.

The EF-111As were an ECM conversion by Grumman Aerospace of the standard F-111A airframe with 18 in service with the 42nd ECS at Upper Heyford. Over 6,000lb of electronic systems including jammers, receivers, computers and antennas were fitted to these aircraft, the most noticeable visual difference being a large fairing on top of the fin which contained a number of antennas, and high-powered jammers were installed in the bomb-bay. In support of the 2nd and 4th ATAFs, the Ravens operated in three different roles. The first was a stand-off jamming with a formation of up to three EF-111As at high altitude disrupting the enemy radars so that all that appears on the screens is a blanket of "snow". This was to protect the slower-flying aircraft such as the AWACS, TR-1As and flight-refuelling tankers which could be vulnerable to attack. The second role was close-in jamming at low level to disrupt the battlefield radars and provide protection for the low-level attacks on ground formations and troops by A-10As, Harriers and Jaguars. This gave protection from radar-guided missiles, but not ground-launched heat-seeking missiles or visual weapons. The third role was the most important task of the EF-111s; escort of the strike aircraft carrying out deep penetration raids over enemy territory. During these operations, the use of the EF-111 terrain-following radar was essential. During the night of 14/15 April 1986 when the Americans launched a night-time attack against Libya, the EF-111As demonstrated the ability to put enemy air defence out of action, while on a deep penetration mission into a potentially heavily defended target area. This allowed the UK-based F-111s to carry out precision strikes on five targets in Libya with much reduced risk, avoiding collateral damage and loss of life to civilians. One F-111F was shot down by anti-aircraft guns, killing the crew of two.

With the ending of the Cold War the F-111s at Upper Heyford continued to be busy, providing support for the raid on Libya in April 1986, in which five of the EC-111As took an active part. The Wing participated in many exercises both in the UK, USA and Europe/Middle East. In the build up to Desert Storm, 42nd ECS EF-111As moved to Taif in Saudi Arabia and by the end of the conflict, the EF-111As had flown 219 sorties day and night from Saudi Arabia, and a further 252 sorties from Incirlik in Turkey. Amongst the targets of the F-111Es of the 20th TFW were airfields, power stations, petrol refineries and suspected nuclear sites, 1,798 sorties being flown without loss. The majority of the F-111Es returned to Upper Heyford on 9 March 1991, and on 6 April the 42nd ECS deployed to Incirlik to help protect the Kurdish minority in Iraq in operation Provide Comfort, and in May the Taif detachment returned to its base.

Following the return from the Gulf War, the crews of the 20th TFW continued attending exercises in Britain and then USA, with detachments to Incirlik and Taif. On 1 October 1991 it was re-designated the 20th FW although the F-111s were definitely strike aircraft, and the squadrons became fighter squadrons. Exercises and training continued into 1992 where the GBU-12 laser guided bombs were dropped, but post-Cold War policy decreed that the F-111s would be withdrawn from Britain by the end of 1993 with Upper Heyford due to close by the end of 1994. Therefore the last EF-111 returned from deployment at Incirlik in June 1992 and the 42nd ECS was deactivated on 10 July, the last EF-111 departing for the USA on 7 August. On 23 April 1993 the 79th FS was the first to deactivate, the last of its aircraft leaving on 10 May, and after participating in Excalibur 93 in June, the 77th FS deactivated on 9 July with its last F-111 departing in August. With only the 55th remaining, six of their aircraft took part in Incirlik Dynamic Guard between 20 September and 8 October, deactivating on 15 October, with the last three F-111s leaving the Upper Heyford runway on 7 December 1993. The runway was closed on 15 December, and final base closure was completed 15 September 1994.

The airfield has been designated a Cold War Monument by English Heritage, but this does not include many of the 1925

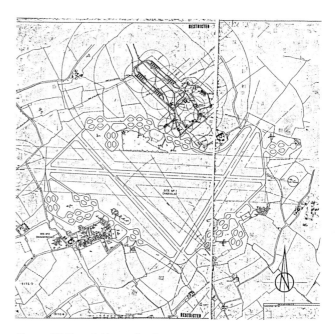

Above: Wethersfield was first intended as a temporary USAAF base, opening in January 1944 for RAF Bomber Command. The 1946 plan shows the three-runway layout with a basic technical site in the southwest corner featuring one T.2 hangar and another T.2 on the southeast boundary. A large weapons storage site was located on the north side of the airfield, which was later adapted for nuclear-weapons storage. Before the RAF could use the station, the USAAF 9th AF arrived in February 1944 for the invasion of Europe and the airfield closed in July 1946, being put under C&M. In June 1951 the USAF returned and the main runway was extended at both ends by 1,000 feet for the operations of the 20th TFW F-84s and later F-100Ds. Wethersfield was put on a standby basis in May 1975 and flying ceased in 1991 when the MOD took over. *RAF Museum*

RAF buildings especially when they have been modified from their original configuration, although it is hoped that buildings like the Station HQ just inside the gate can be retained. The facilities to be preserved include the 56 HAS, especially the Quick Response Area, where aircraft were kept at readiness in eight HAS within a dedicated compound complete with living accommodation for the crews on standby and ready to depart at short notice. The weapons storage site, which is surrounded by a double security fence is included in the listing as well as the fenced off ATC, runways, perimeter tracks dispersals and aprons on this vast site. It is possible that the RAF 'A' Type hangars may not be retained, as they do not represent the Cold War, but the early RAF Expansion Period. The airfield site is now occupied by companies mainly in the motor trade and is used for testing new models before they are launched, so security is tight. The 1925 RAF domestic site over the road from the airfield is not listed and there are plans to develop the area for just over 1,000 houses, in addition to the ex-USAFE housing which is now occupied by civilians.

Wethersfield, Essex;
TL720335, SE of Finchingfield by B1053

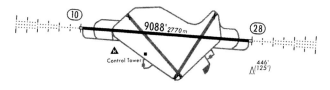

Wethersfield was built as a satellite to Ridgewell as a temporary USAAF bomber base with three concrete runways, the longest being east-west at 6,000 feet and a perimeter track with fifty dispersals. The basic technical site was located on the SW corner and included two T.2 hangars with most other accommodation in Nissen huts. The station was declared completed on 30 June 1943 and opened as a full RAF Bomber Command station in January 1944. It was passed to the US 9th AF on 1 February as part of the Allied build up in preparation for the invasion of Europe and the first residents were the A-20G Havocs of the 416th BG. The Havocs began tactical daylight bombing on 4March where pre-invasion targets and V-1 launch sites were hit. With the advances of the Allies after D-Day, the targets were too far from the base and the 416th moved to advanced airfields in France. The transfer of 38 Group RAF Stirlings transports began on 9 October 1944, but both 196 and 299 Squadrons were not fully equipped until November. The squadrons undertook supply drops to resistance groups and glider tug training, but moved to Shepherd's Grove by the end of January 1945. The 9th AF returned with 81 C-47s to carry the 6th Airborne Division on the Rhine Crossing on 24 March 1945 after which the Americans departed and the base returned to the RAF. Wethersfield was used for flying training from October 1945 until July 1946 when the station closed and was reduced to C&M.

In September 1950, Wethersfield was chosen as a postwar fighter base and on 16 April 1951 Fighter Command took control to investigate what would be required to bring the neglected base back to operational use. On 30 June 1951 it was officially allocated to the USAF and additional land was acquired to lengthen the main runway by 1,000 feet at both ends,

Above: Wethersfield had three runways constructed with the longest main runway at 6,000 feet and 50 dispersals were located around the perimeter track. When the USAFE arrived in June 1951, the main runway was extended to 8,000 feet with 42 dispersals upgraded for 84 fighter-bombers. By May 1969 the main runway had been further extended to 9,088 feet as shown on the approach chart, but operational flying ceased in June 1970, and it is now an MOD police training facility. *Jeppesen*

while 42 hardstandings were upgraded to accommodate up to 84 fighter-bombers. Amongst other improvements included a new bulk fuel store, and runway and taxiway edge lighting. The USAF 3rd AF took control on 24 August 1951 with rebuilding which included a nuclear-weapon store on the north side of the base. Accommodation was generally temporary with prefabricated huts, and the perimeter was secured by a 10ft high, 3,500ft long chain-link fence with barbed-wire topping.

On 1 December 1951 the base was allocated to the 49th Air Division who declared Wethersfield operational on 12 February 1952. The 20th FBG began to move in from 22 May, followed by two squadrons of 50 F-84G Thunderjets which flew across the North Atlantic via Labrador and Iceland, and were capable of carrying small tactical nuclear weapons. The 20th FBG consisted of the 55th, 77th and 79th FBS, the latter being based at Woodbridge and overall control passed from the 49th AD to HQ 3rd AD on 5 June 1952. Training was in general and dive bombing techniques in Britain, Europe and North Africa, the 20th becoming a FBW in February 1955. In June 1955 the F-84Gs began to be replaced by the swept wing F-84F Thunderstreaks with a much greater weapons carrying capability, but began to be replaced by the supersonic F-100D and F-100F Super Sabres from June 1957, becoming the 20th TFW on 8 July 1958. Continuing improvements were made to the base with extra

Left: The 20th TFW re-equipped with F-100D Super Sabres at Wethersfield from June 1957, and continued until replaced by F-111Fs in June 1970. F-100D 63003 of the 20th TFW is seen taking off from Wethersfield. *Michael Bowyer*

Left: The 20th TFW began to receive the F-100Ds Super Sabres in June 1957 and 0-62986, fitted with underwing drop tank and still in natural metal finish, was at a Wethersfield open house soon after the type arrived. *Author*

accommodation built in 1967. From 1966 the colourfully-marked metal-finish F-100s were camouflaged, but with the arrival of the F-111s, the 55th TFS departed on 1 May 1970, followed by the 77th on 1 June, when the 20th TFW moved to Upper Heyford.

Wethersfield became a standby base and between May and August 1975, the 349th SRS, 100 SRW flew five highly secret U-2C missions from there, but after one of the Wethersfield-based aircraft crashed at the German ski resort on Winterberg on 29 May 1975, the secret was revealed. After these intelligence-gathering aircraft departed, the base was reduced considerably. The final flying activity was by RAF Phantoms from Wattisham while their runway was being repaired and the USAF moved out in the spring of 1991, since when the base became a depot and training facility for MOD police.

Woodbridge, Suffolk;
TM330487, 11 miles NE of Ipswich, off B1084.

The main role of Woodbridge was as one of three long, wide runways to accept Allied bombers in emergency during World War 2, the other two being Manston in Kent and Carnaby in Yorkshire. With the increased bomber offensive in 1942, many of the crippled bombers were unable to find a safe haven to land when badly damaged, as most of the available airfields were inadequately equipped, resulting in the three runways being constructed close to the East Coast, 9,000ft long and 750ft wide with overshoots at either end of 1,500 feet, giving an overall length of 12,000 feet. Each runway was divided into three lanes. The southern lane with green lights at night and white lights by day was the emergency lane on which any aircraft in distress could land without contact with air traffic, in case the radios were out of action. The centre lane was illuminated by white lights at night and the northern one with yellow, clearly defining the landing areas, with approach lighting to guide the pilots. The middle and northern runways would be controlled by air traffic, advising of any obstructions to avoid caused by preceding emergencies. In the event of poor visibility, FIDO was installed with petrol-filled jets alongside the runway, which was effective in dispersing fog. The Woodbridge runway was located in the middle of Forestry Commission land, resulting in the destruction of over a million small trees when construction started in July 1942. Even before completion a B-17 made an emergency landing on 18 July 1943 when it was short of fuel. Woodbridge opened

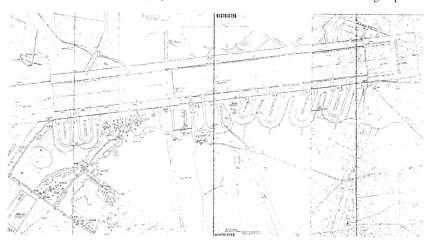

Above: Woodbridge featured a 9,000ft long by 750ft wide three-lane runway for World War 2 emergency landings to avoid blocking home base runways with returning aircraft in the event of a crash landing. There were 1,500ft overshoots at each end and the small basic technical site had a single B.1 hangar located on the south side of the runway. There were also six emergency loops on the south side of the runway where damaged aircraft could be dragged off the runway to make room for more landings. The airfield was abandoned by the RAF in March 1948, but transferred to the USAF in June 1952 for the 20th FBW, which operated F-84s, F-101s, F-4Cs and finally A-10As as twin base to Bentwaters. The airfield closed to operational flying in August 1992, and is now occupied by the Army. *RAF Museum*

Left: The first F-4D Phantom arrived at Woodbridge on 23 April 1969 for the 81st TFW, 67676 taxiing in at the NATO Tiger Meet on 7 August 1969 with the background of Forestry Commission conifer woodland. *Author*

on 1 September 1943 and the runway was operational from 15 September, with 36 emergencies arriving in the first two weeks, and a total of 4,120 emergency landings being made by the end of June 1945. As an additional aid to bombers in poor visibility, at the end of April 1944, a line of sodium lights was installed along the south side of the runway extending east to within a mile of the coast at Orford. The station was closed for five days on 19 March 1945 to allow two Halifax squadrons towing 68 Horsa and the larger Hamilcar gliders as part of the Allied airborne crossing of the Rhine.

Following the end of World War 2, the station was used occasionally by experimental aircraft from Martlesham Heath, but the site was abandoned by the RAF on 15 March 1948 as being surplus to requirements. RAF interest returned on 26 May 1952 when Woodbridge came under the control of Fighter Command. It was transferred to the USAF on 5 June 1952 and the first residents were the F-84Gs of the 79th BS of the 20th BW, which were later replaced by F-84Fs and from 1957, F-100s. From 1953 improvements were made to the base, with a new squadron operations building, a new fire station in 1956 and housing accommodation ready for occupation from April 1957. Control was taken over by the 81st TFW at Bentwaters on 8 July 1958 when the two airfields were twinned and preparations were made for a second squadron although the facilities were somewhat basic. The runways were resurfaced in December, with additional turn-offs to reduce taxiing and in January 1959 the 78th TFS arrived from Shepherd's Grove, converting from F-84Fs to F-101s. The first F-4C Phantoms arrived on 2 March 1966 for the 81st TFW, the 78th TFS starting conversion on 23 April, and was fully re-equipped by July. In November 1969 the 79th TFS moved to Upper Heyford and in December and January 1970 the 67th Aerospace Rescue and Recovery Squadron arrived from Spain with HC-130Ns and Ps, and HH-3E Jolly Green Giant helicopters.

As part of the expansion of the 81st TFW with the A-10A tank busting aircraft, four squadrons were formed with the 91st TFS moving from Bentwaters to Woodbridge in 1978. HAS were constructed on the south side of the runway at the eastern end of the airfield to protect the aircraft from conventional attack. On 15 May 1992 the 78th TFS deactivated at Bentwaters, followed by the 91st TFS on 28 August and all operational flying ceased at Woodbridge, apart from some transport flights removing equipment. The base was taken over by the Army two years later and has been used occasionally for flying training with Lynx and Gazelle helicopters. The station has been developed by the Army with new buildings constructed on the original modest technical site on the south side of the runway.

Below: For the NATO Tiger Meet in August 1969 F-100Ds of 79th TFS, 20th TFW visited Woodbridge including 0-62997 with 0-62977 beyond. The heat of the jet pipes blistered the camouflage paint around the rear fuselage. In the background is a Nissen hut. *Author*

Chapter 6

RAF Strategic Fighter Stations

Binbrook, Lincs;
TA190960, 9 miles SW Grimsby, by B1203

With the transfer from Bomber Command to Fighter Command in April 1960, a number of improvements were made to accommodate the new jet fighters entering service with the RAF, including the lengthening of the main runway to 7,500 feet, the work being completed in time for the reopening on 1 June 1962. The first flying unit to arrive was 64 Squadron with Javelin FAW.9s which remained until 1 April 1965, when the squadron departed for Singapore. The operational Javelins were joined by the Central Fighter Establishment (CFE) in October 1960 equipped with Hunters, Javelins and Lightnings with the task of developing the latest fighter tactics. In May 1963 85 Squadron returned, but for aerial target duties operating some of the last Meteor F.8s and Canberras, including the radar-equipped T.19.

With the departure of 64 Squadron, the station prepared for the arrival of the Mach 2 Lightning F.6 which equipped 5 Squadron when it re-formed on 8 October 1965. In February 1966 the CFE disbanded, but part of it formed the Air Fighting Development Squadron (AFDS) which became the Fighter Command Trials Unit until it disbanded on 30 June 1967. In January 1972 85 Squadron moved to West Raynham, making way for 11 Squadron, which moved in from Leuchars on 28 March with Lightning F.6s. The main task of the Lightning squadrons was to intercept potential Soviet raiders at high altitude on the approaches to Britain, and maintained two aircraft in a Quick Reaction Alert (QRA) hangar, known as the Q-Shed, with the pilots and aircraft ready for a rapid scramble when a potentially hostile unidentified aircraft was detected. Because the endurance of the Lightning was fairly short, flight-refuelling tankers were often available to keep the aircraft in the air longer. With the longer-endurance Phantoms entering service with the RAF in 1974 and taking over the air defence role, withdrawal of the Lightnings commenced, leaving Binbrook the last operational base of the type, and adding the responsibility for the conversion and operational training, which from September 1974 was the Lightning Conversion Unit. As the training duties reduced, it became the Lightning Training Flight, operating some Lightning T.5s. With the final withdrawal of the Lightning from RAF service, 5 Squadron disbanded in December 1987 to re-equip with Tornado F.3s at Coningsby, followed by 11 Squadron in June 1988, after which the station closed and has been sold for development. Most of the buildings remain, the technical site used for industrial purposes while the married quarters have been developed as a housing estate in the new village of Brookenby, but the runways and perimeter tracks have mostly been removed.

Below: No 85 Squadron operated an assortment of Canberras and Meteors at Binbrook from 25 April until 28 January 1972 when it moved to West Raynham. The squadron flew Canberra T.19s on target facilities duties from August 1965 until disbandment at West Raynham in December 1975. Canberra T.19 T:85 Squadron is on dispersal at Binbrook on 16 October 1969. *Author*

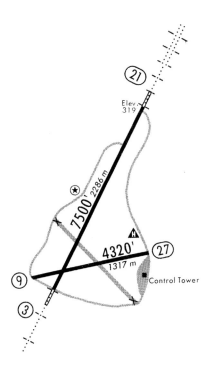

Left: Binbrook opened in June 1940 with five austerity 'C' Type hangars as part of the technical site located on the eastern boundary of the airfield within 1 Group RAF Bomber Command. The airfield was closed in September 1942 for the construction of three concrete runways with interconnecting perimeter track and bomber dispersals. Binbrook reopened in May 1943 with RAAF Lancaster squadrons, which continued until the end of World War 2, with the last mission in April 1945. The station continued in the strategic bomber role when Lincoln B.2s arrived in July 1946, and were withdrawn when replaced by Canberras from June 1951 when 101 Squadron was responsible for the introduction of the new jet bomber. The Canberras left in December 1959, when the station was put under C&M in preparation for the arrival of Lightnings, the transfer to Fighter Command being made in April 1960. As part of the improvements the main 21-03 runway was extended to 7,500 feet and reopened in June 1962. The Lightnings of 5 Squadron arrived in October 1965 followed by 11 Squadron, and remained until the Lightnings were withdrawn in June 1988, when the RAF station was closed. The March 1968 chart of the airfield shows both the main runways and runway 09 – 27 still in use, but the shorter runway was later disused. Although the majority of the technical site remains in commercial use, the runways have been removed. *Jeppesen & Co*

Coltishall, Norfolk;
TG270225 9 miles N of Norwich near B1150.

Extensive improvements were made to the station during 1958 in preparation for the arrival of Lightnings, and on 1 September 1959 the Air Fighting Development Squadron (AFDS) arrived, the first RAF unit to fly the Lightnings. The AFDS moved out to Binbrook on 5 October 1962 and the Lightning Conversion Squadron operated from Coltishall from January 1960 until departing in August 1961. On 2 August 1960 74 Squadron received its first Lightning F.1 and operated from Coltishall until February 1964 when it moved to Leuchars, to be replaced by 226 OCU, the Lightning Operational Conversion Unit marked with the colours of 145 (Reserve) Squadron, which was the centre of Lightning training until September 1974, when the Jaguars arrived.

The first Jaguars to arrive was 54 Squadron in August 1974 with 6 Squadron in November and 41 Squadron, specialising in tactical reconnaissance duties, arriving in April 1977 to form a very effective mobile reaction force. The Jaguar was a single seat fighter-bomber with low maintenance demands and ease of deployment, and the fleet was allocated to the NATO Rapid Reaction Force. The Anglo-French Jaguar was produced jointly by Breguet and the British Aircraft Corporation (BAC) in a company known as SEPECAT which was registered in France in May 1966, and the aircraft was powered by two Rolls-Royce/Turbomeca Adour turbojets, under a memorandum of understanding signed by the two governments on 17 May 1965. The initial agreement called for 150 aircraft for each country, but early in 1967 the total for each country was increased to 200, with the RAF taking 90 strike versions and the remainder two-seat advanced trainers. The single-seat version for the RAF was designated the GR.1 and was armed with two 30mm Aden cannons, could carry weapons on five external stores points and was fitted with an inertial nav/attack system, laser rangefinder and head-up display (HUD). The RAF two-seat version was designated the T.2; it was armed with one 30mm cannon, had provision for five weapons points and was fitted with the same nav/attack system as the single-seat version. By 1970, the RAF requirements had been further revised, the overall 200 aircraft consisting of 165 single-seat strike versions and only 35 two-seat trainers.

Jaguars had been designed from the start to make short take-offs from rough fields, with reduced maintenance costs and a flexibility in the choice of weapon systems, while retaining a high development potential. A typical tactical weapons load was four 1,000lb bombs carried externally with a typical turnaround time of 15 minutes. The weapons choices included air-to-surface guided missiles, air-to-air guided missiles of the Sidewinder type, two 30mm cannons and all types of unguided bombs and rockets, together with reconnaissance pods and external fuel tanks. Following the signing of the Anglo-French production agreement on 9 January 1968, the first French prototype made its maiden flight from Istres on 8 September. The first British assembled prototype was rolled out from the BAC factory at Warton on 18 August 1969, and made its maiden flight on 12 October. It was not until 12 October 1972 that the first production Jaguar for the RAF made its maiden flight from Warton, followed on 16 November with the second aircraft which was the first to be fitted with a laser ranger and target marker, giving a high weapons delivery accuracy. Data

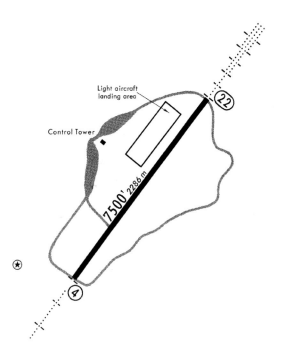

Left: Coltishall was built as a late RAF Expansion Scheme bomber station, but with reduced specification 'C' Type hangars and basic administration, technical and living accommodation. The station opened as part of 12 Group Fighter Command and retained the fighter and strike role throughout its existence, specialising in the night fighter task after World War 2. Vampire and Meteor night fighters were introduced from late 1951, later replaced by Javelins. During 1958 improvements were made, including the extension of the runway to the southwest to a total of 7,500 feet ready for the first Lightnings to arrive in September 1959. The Lightnings were replaced by Jaguars from August 1974 and continued as the main UK RAF Jaguar base until they were withdrawn and the station closed in April 2006. The May 1968 airfield chart shows the fully extended main runway and the substantial ASPs on either side of the air traffic control with the hangars located beyond. *Jeppesen & Co*

supplied to the on-board computer would guide the aircraft and automatically release the pre-selected weapons at the target, following an automatic departure from the target zone, ready for an attack on another target, or return to base.

Although the first delivery to the RAF was on 30 May 1973 to Lossiemouth in Scotland, this was to be the main training base and where the squadrons were formed initially, while Coltishall was to be the main UK operating station. With pilot conversion training commencing in October 1973, with 54 Squadron becoming the first RAF unit at Lossiemouth on 23 April 1974, and after work-up moved to Coltishall on 15 August. The second unit was 6 Squadron, which was the first to have Jaguars fitted with the laser nose, and it re-formed at Lossiemouth on 1 October 1974, moving to Coltishall on 15 November. By now the emphasis at Lossiemouth was on the training and formation of additional squadrons, with five Jaguar units based in Germany as part of the 2nd TAF, their task being interdiction, battlefield support and reconnaissance, while the Coltishall-based units were to form part of a mobile force in conjunction with the Third Division of the British Army. In the spring of 1975 54 Squadron was declared combat ready, soon to be followed by 6 Squadron, while at the same time the first Jaguar squadrons were re-equipping in Germany. The role of the Coltishall-based squadrons was to support ground forces in Europe against a Soviet attack, initially with conventional weapons, but also the WE177 tactical nuclear weapon, eight of which were issued to each squadron before they were finally withdrawn in 1994.

A specially-designed reconnaissance pod was developed with flight trials commencing in the spring of 1975 to allow low-level reconnaissance and speeds approaching Mach 1, the pod being mounted beneath the fuselage centreline; the two specialist units were II Squadron at Laarbruch in Germany and 41 Squadron at Coltishall, which re-formed in its specialist role on 1 April 1977. To build on their training skills, pilots of 6 Squadron took part in Exercise Maple Leaf in Canada in 1980 and 54 Squadron participated in the demanding Exercise Red Flag at Nellis AFB, both with a high level of success.

Following the end of the Cold War,

Left: No 41 Squadron re-formed at Coltishall on 1 April 1977, specialising in tactical reconnaissance. Jaguar GR.1A XZ363 FO:41 Squadron is seen leaving the Coltishall runway on 6 July 1998. *Author*

Above: No 74 – 'The Tigers' Squadron was the first RAF squadron to re-equip with the Lightning F.1s, having moved to Coltishall on 8 June 1959, and began to receive its aircraft a year later. The Lightning F.1s were armed with de Havilland Firestreak infra-red heat seeking air-to-air missiles. BAC

the Coltishall Jaguar squadrons participated in operations in northern Iraq and the Balkans, but despite being a great success and relatively inexpensive to operate, premature withdrawal commenced with the disbanding of 54 Squadron on 11 March 2005, and with the closing of Coltishall on 1 April 2006, 41 Squadron disbanded and 6 Squadron departed to Coningsby until it also disbanded on 31 May 2007. Coltishall was stripped of everything useful with the domestic houses sold, and the technical site to be used as a prison.

Coningsby, Lincs;
TF225565, south of Coningsby village by A153

With the departure of the Vulcan V-bombers to Cottesmore in November 1964, preparations commenced to make Coningsby available for the revolutionary TSR-2 strike bomber, but with the British government's cancellation of this advanced aircraft on 6 April 1965, Coningsby was reduced to C&M until 1966, when it was selected for the introduction of the Phantom FGR.2s to the RAF.

Following on from the cancellation of the TSR-2, an anglicised version of the General Dynamics swing-wing F-111 was planned, but with the cancellation of that programme, the RAF was without a modern strike aircraft, and the Phantom appeared to be the answer, since the RAF was relying at the time on the Hunter. To provide some work for the British aerospace industry, the RAF FGR.2s and the FG.1s for the Fleet Air Arm (FAA) were to be anglicised by about 50 percent of value, the major change being the replacement of the GE J-79 turbojet by the Rolls-Royce Spey turbofan engine, with production authorisation for the RAF F-4M being given on 5 May 1965. The principal airframe parts made in Britain were the outer wing, complete with droop flaps, made by Shorts in Belfast and the complete rear fuselage and tail, fabricated by BAC in Preston. Ferranti worked with Westinghouse on the radar fire-control system, as well as supplying a new inertia nav/attack system. To support British Phantoms in service, Hawker Siddeley Aviation (HSA) at Brough were appointed sister company to McDonnell with full design authority for major servicing, repairs and modifications, with flight testing based at Holme-on-Spalding-Moor.

The first development YF-4M made its maiden flight from St Louis on 17 February 1967, and the first to be delivered to the RAF was ferried to Aldergrove in Northern Ireland on 20 July 1968, arriving at Coningsby for 228 OCU on 23 August. The primary task of the RAF Phantoms was close support, ground attack and tactical reconnaissance mostly at low level. The training task was given to 228 OCU which re-formed in October 1968 and carried the markings of 64 (Shadow) Squadron, as in times of high tension, the squadron was available for action flown by the instructors and advanced students. The initial crews trained by 228 OCU were allocated to be the nucleus of 6 Squadron which re-formed also at Coningsby on 7 May 1969 as the first RAF front line Phantom FGR.2 unit and operated the aircraft until converting to Jaguars at Lossiemouth on 1 October 1974. Other units which operated Phantoms from Coningsby included 29, 54 and 111 Squadrons. No. 54 Squadron re-equipped on 1 September 1969, remaining at Coningsby until disbandment on 22 April 1974 when it converted to Jaguars. On 1 April 1972 41 Squadron re-formed in the strike/reconnaissance role and carried forward and oblique cameras, sideways-looking aircraft radar (SLAR) and IR Linescan. No. 41 Squadron remained at Coningsby until it was disbanded on 31 March 1977, and re-formed the next day with Jaguars at Coltishall. On 1 October 1974 111 re-formed, but on 3 November 1975 moved to Leuchars to operate the ex FAA Phantom FG.1s in the air defence role. The final Phantom unit was 29 Squadron which

Left: By June 1982 the Phantom force was adopting an overall grey as shown on Phantom FGR.2 XV401 B:64 Squadron/228 OCU based at Coningsby, but visiting Brize Norton on 11 June 1982. *Author*

re-formed on 1 January 1975 and was the last FGR.2 unit when it disbanded on 30 March 1987, reforming the next day with Tornado F.3s.

With the departure of the Phantoms, Coningsby was prepared for its next residents, which were the air defence version of the Tornado. In addition to the two 'J' Type hangars, one B.1 for use during World War 2, a 'Gaydon' hangar had been built in 1960 and three T.2s were also in use, and a batch of HAS were constructed to protect the aircraft from conventional attack. The RAF was tasked with maintaining the integrity of Britain's Air Defence Region (UK ADR), which is a massive area extending from Iceland to the English Channel and from the Atlantic to the Baltic approaches. This is significantly larger than NATO's Central Region where the threat was from small agile fighters at short range. For Britain, the primary threat was from large formations of Soviet strategic bombers, such as the Tu-22M, which could carry nuclear-armed cruise missiles south from the North Cape. To defend against this threat the RAF needed a long-endurance Beyond Visual Range (BVR) missile-armed interceptor which could loiter, with the help of in-flight refuelling, up to 400 miles from base and able to patrol the entire UK ADR. It would have to be able to detect and engage a number of targets in all weathers from a stand-off range using a powerful radar, together with information from the patrolling AWACS aircraft. With the growing threat of Soviet and Warsaw Pact air power in the 1970s, the Lightning and later the Phantom were unable to cope with the new threat. The result was the launch of the Tornado ADV, with full-scale development started on 4 March 1976 with a requirement for 165 aircraft, the aircraft having 80 percent commonality with the Tornado IDS. Armament was four medium-range BAe Skyflash missiles adapted from the American AIM-7 Sparrow recessed under the fuselage, with a new fire-control Foxhunter radar in the nose under an extended and more pointed radome. A single 27mm Mauser cannon was retained on the starboard side with a retractable flight-refuelling probe on the upper port side of the nose. The electronic systems were enhanced considerably to cope with the growing threat and the crew of two consisted of a pilot in front and navigator behind with the shared task of weapons-systems operators. Although the Tornado ADV was not a dogfight aircraft, as all engagements were planned BVR, in case of the need for close-in defence two AIM-9 Sidewinders were carried, on underwing pylons which pivoted as the wing sweep angle varied. These interim weapons were later replaced by the American AMRAAM and ASRAAM air-to-air missiles and the first Tornado F.3

Below: Tornado F.3 ZH555 CV:5 Squadron with the flight refuelling probe extended ready to take on fuel from a TriStar tanker. *Author*

Above: No 6 Squadron was the first frontline unit to equip with the Phantom FGR.2 on 1 May 1969 and continued to operate the type until disbanding on 30 September 1974 to convert to Jaguars. Phantom FGR.2 XV456 J:6 Squadron is on the Coningsby flight line. *Author*

was delivered to Coningsby on 28 July 1986 for the training of the students for 29 Squadron.

The Tornado F.3 crew had to be able to react quickly to any threat, with extremely flexible air defence operations capable of meeting the many types of threat, as well as dealing with the element of surprise expected in war. The squadron crews were capable of carrying out interceptions at all heights and speeds with the aircraft and systems virtually an extension of the pilot and navigator, and they had to be competent to operate the aircraft in all weathers and by day and night. The Skyflash BVR AAM was fitted with an advanced radar-proximity fuse, with the warhead capable of destroying both supersonic and subsonic targets from low level to the higher altitudes in a single shot. The Tornado F.3s often operated in pairs, approaching a hostile formation from both sides. As already mentioned, the Tornado F.3 was not a dogfighter, but a long-range bomber destroyer, the attack being by the BVR mode using hit-and-run tactics. In the case when fighters were targets, the pair of Tornados approached the targets rapidly and engaged at the greatest possible range before breaking away.

To undertake the training of air and ground crews, 229 OCU was re-formed on 1 November 1984 carrying the shadow markings of 65 Squadron, later on carrying the markings of 56 Squadron from 1992 with twelve dual-control aircraft and twelve of the normal single-stick configuration to allow training of the navigators. The first operational unit was 29 Squadron, which re-formed at Coningsby on 1 April 1987, followed by 5 Squadron, which replaced its Lightnings with Tornado F.3s on 1 May 1988. Both squadrons formed a combined unit for Operation Desert Storm and were the first to deploy in Saudi Arabia in August 1990. No. 29 Squadron disbanded in 1998 and now is equipped with the Typhoon, and 5 Squadron disbanded in 2003, to reform at Waddington on 1 April 2004 to prepare for the operation of the Sentinel.

RAF Coningsby is now the busy main operating base for the RAF Typhoons in the air defence and ground-attack role, and in contrast, is also the home of the Battle of Britain Memorial Flight with Hurricanes, Spitfires and the Lancaster.

Leeming, N Yorks;
*SE305890, by A1 Great North Road,
7 miles SW of Northallerton*

The role of Leeming changed from flying training to air defence in 1988 when preparations were made for the operation of three Tornado F.3 squadrons. The duties of these squadrons was Quick Reaction Alert (QRA) South and the first to reform was 11 Squadron in August 1988, which was armed with the ALARM anti-radiation missile from February 2003 for the suppression of enemy air defences (SEAD) complimenting the Tornado GR.4 force. Following 11 Squadron on 1 November 1988 was 23 Squadron, and 25 Squadron on 1 October 1989. Although the Cold War was over, 25 Squadron intercepted eight Tu-95 Bear-Hs and two Tu-160 Blackjack Russian strategic bombers in July 2007, escorting them through

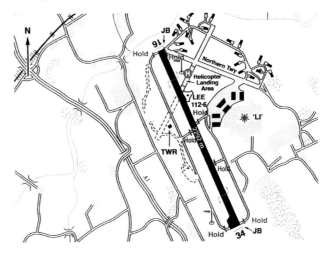

Above: At the time of writing Leeming is still an active RAF station. It was constructed as an RAF bomber station with five austerity 'C' Type hangars in the southeast-located technical site and opened for flying in June 1940. Three runways were constructed in time for RCAF heavy bomber operations to start in May 1943, and in 1965 the main 34/16 runway was extended southeast to allow fast jet operations. With the departure of the Javelins, the station was taken over by Training Command until August 1988 when the first of three Air Defence Tornado F.3 squadrons arrived to maintain the southern QRA. During this period 21 HASs were constructed on the northern boundary and Tornado operations continued until the disbandment of 25 Squadron in April 2008. The station is now the base for 100 Squadron with Hawks used for air defence training.
Pooley's Flight Guides

UK airspace. With the reduction in the overall threat, 23 Squadron disbanded on 26 February 1994 to operate AWACS from Waddington jointly with 8 Squadron. In October 2005 11 Squadron had its Tornados withdrawn and it moved to Coningsby to operate the Typhoons. The last air-defence unit at Leeming was 25 Squadron, which disbanded on 4 April 2008 when the QRA was relocated to Leuchars, together with the aircraft. Leeming remains an operational RAF station with the Hawks of 100 Squadron used for air combat training.

Leuchars, Fife;
NO460205, 5 miles NW of St Andrews, by A919.

From the start of the Cold War, Leuchars has been at the forefront of the air defence of the UK with responsibility for the QRA North, covering the northern approaches to Britain across to Iceland. The third-generation jet fighters first arrived at Leuchars in the form of Lightning F.1s when 74 Squadron moved from Coltishall on 2 March 1964, followed by 23 Squadron replacing its Javelins with Lightning F.3s in August. At last the RAF operated an air-defence interceptor capable of defending against Soviet high-flying strategic bombers. No. 74 Squadron re-equipped with Lightning F.6s in September 1966 before flying to Tengah in Singapore on 12 June 1967, their replacement being 11 Squadron, which re-formed with Lightning F.6s on 3 April 1967. The squadrons took turns to man the QRA ready for rapid scramble when potential hostile aircraft were detected approaching the UK, but sometimes these were civil aircraft which had failed to communicate their intentions. When Soviet aircraft were intercepted, they would be graphed to check for any modifications or updates detectable externally. The replacement of Lightnings at Leuchars began with the reforming of 43 Squadron with the Phantom FG.1s on 1 September 1969, this version being surplus to FAA requirements, and allocated to air defence duties with the RAF. No 11 Squadron left for Binbrook on 28 March 1972, and the replacement of the Lightnings was completed when 23 Squadron disbanded on 31 October 1975 to prepare for re-equipping with Phantom FGR.2s at Coningsby, with 111 Squadron arriving from Coningsby on 3 November with Phantom FGR.2s. These strike versions of the Phantom were replaced by air-defence FG.1s in January 1979 when additional aircraft came available from 892 NAS disbandment on 15 December 1978.

The Fleet Air Arm (FAA) version of the F-4K Phantom was fitted with similar systems in most respects to the RAF F-4M, but to accommodate its overall length limitations on Royal Navy (RN) carriers of 54 feet, the entire nose-mounted radar installation with radome was hinged to fold back 180 degrees. To allow full-weight take-offs from HMS *Ark Royal*'s less powerful catapults, the undercarriage nose leg was extended by 40 inches/1.03m, allowing a reduced wind-over-deck (WOD) speed. To further help with the catapult take-offs, larger wing flaps were fitted, drooped ailerons and inverted slats on the tailplane leading edges. Many of these improvements were incorporated in the newer American F-4J. To permit higher vertical velocity and heavier weight landings, the strengthened main undercarriage and larger tyres of the F-4J were used, as well as a stronger arrester hook. An all-British navigation and communications systems were fitted and the cockpit instrumentation was extensively revised. The first three of 28 FAA Phantom FG.1s were ferried to Yeovilton on 25 April

Below: Leuchars-based No 43 Squadron (The Fighting Cocks) re-equipped with surplus FAA air defence Phantom FG.1s in September 1969, later replacing them with Phantom FGR.2s. Phantom FG.1 XT875 K:43 Squadron was one of the earlier versions, and was visiting Chivenor on 22 August 1970. *Author*

Left: A trio of thirsty Leuchars-based 43 Squadron Phantom FGR.2s, with XV581:AE closest, await their turn for taking on fuel from 216 Squadron TriStar ZD953 on 10 May 1988. *Author*

1968, which was to be the initial base, these aircraft being sufficient for two squadrons, but by that time the decision had been made to withdraw British forces from east of Suez, only 892 Squadron was to be formed, supported by 767 NAS for conversion training. 20 of the surplus FG.1s were allocated to the RAF for 43 Squadron, and to bring the fleet together, the Phantom Training Flight moved to Leuchars in 1972, to be followed by 892 Squadron in the same year. When the 892 Squadron Phantoms were not at sea on HMS *Ark Royal*, they shared the QRA with 23 Squadron Lightnings and 43 Squadron Phantoms.

All the RAF air defence Phantom units spent a month annually at the Armament Practice Camp (APC) in Akrotiri, Cyprus, usually flying there non-stop with the benefit of air-to-air refuelling. This allowed visual interception and live firing of the gun pod against banners towed by Canberras, but air-to-air missiles (AAM) were not fired due to the expense of replacements, although simulated missile attacks were possible in the UK as they were purely electronic simulation. The aircraft carried dark blue inert Sparrows as ballast in the under-fuselage recesses to maintain the correct centre of gravity (C of G), or for training a simulator plug was fitted to provide the same system inputs as a real Sparrow or Sky Flash. In addition it was normal to fly with Sidewinder training rounds with a real IR homing head, allowing a full training interception with the homing head locked-on to a target, without releasing the training round.

Replacement of the air-defence Phantoms began with Tornado F.3s at Leuchars in September 1989, the new aircraft being used in the Gulf War of 1991. No. 43 Squadron re-equipped with Tornado F.3s in September 1989 and shared the northern QRA with 111 Squadron as well as 43 Squadron being responsible for the Falklands QRA with 1435 Flight until the squadron disbanded on 13 July 2009. No. 111 Squadron began to replace its Phantoms with Tornado F.3s from January 1990 and has continued to maintain the Northern QRA until the final retirement of the Tornado F.3 in March 2011. Meanwhile in September 2010 6 Squadron re-formed with Typhoons at Leuchars to take responsibility for the Northern QRA and is likely to be joined by a second Typhoon unit, which has yet to be formally identified. Although the Cold War is long over, the Russian threat still exists with more modern strategic bombers, but the possibility of nuclear holocaust has passed. Following the Government Strategic Defence Review in October 2010 when both Kinloss and possibly Lossiemouth were due to close, attempts are being made to retain the latter due to economic considerations, and instead close Leuchars with the two Typhoon squadrons joining one remaining Tornado unit at Lossiemouth. This would leave just one RAF station remaining in Scotland.

Right: Leuchars first opened in 1918 as an RNAS station with Belfast hangars. In the early part of World War 2 the station expanded with an enlarged technical site including four utility 'C' Type hangars in the NW corner. To allow all-weather operations two concrete runways were built in 1941. Meteor jet fighters arrived in May 1950 followed by Hunters and Javelins and the main runway was extended east in 1956. The station became the northern QRA with Lightnings from March 1964, followed by Phantoms in September 1969 and Tornado F.3s in September 1989. During this period 22 HAS were built at the eastern end of the airfield. No 6 Squadron started to take over the northern QRA in September 2010 and 111 Squadron, the last RAF unit to operate the Tornado F.3s, disbanded on 22 March 2011. *Pooley's Flight Guides*

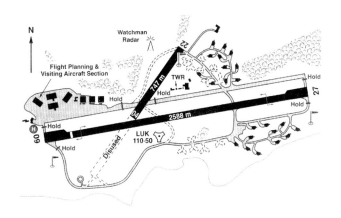

The second Lightning unit to form at Wattisham was 111 Squadron initially with F.1As, but later receiving F.3s. Firestreak armed Lightning F.3 XR716 F:111 Squadron departing Wattisham on 24 April 1968. *Author*

Wattisham, Suffolk;
TM025510, 6 miles SW of Needham Market, by B1078

With the arrival of Lightnings, Wattisham became a major strategic air defence station with, at last, the capability of intercepting hostile raiders. To facilitate rapid departures a QRA was constructed close to the main runway, which housed a pair of Lightnings on readiness for rapid scramble against unidentified targets. Although by this time potential raiders would attempt to approach Britain at low level below the radar, the Soviet strategic bombers and reconnaissance aircraft flew down from the Murmansk area between Iceland and Scandinavia at high level, often testing the UK defences and reaction times.

No 56 Squadron began to exchange its Hunters for Lightning F.1As in December 1960, which were gradually replaced by the improved F.3s in March 1965. Meanwhile, 111 Squadron began their conversion from Hunters to Lightning F.1As in April 1961, which were exchanged for F.3s from December 1964. On 11 April 1967, 56 Squadron was posted to Akrotiri in Cyprus where they converted to the Lightning F.6 in September 1971. The squadron returned to Wattisham on 21 January 1975 and disbanding on 28 June 1976, reforming the next day at Coningsby with Phantom FGR.2s before returning to Wattisham on 9 July. With the departure of 56 Squadron to Akrotiri, 29 Squadron re-formed at Wattisham on 10 May 1967 with Lightning F.3s which it retained until disbandment on 31 December 1974, being re-formed the following day at Coningsby with Phantom FGR.2s. No. 111 Squadron added Lightning F.6s to its F.3s in May 1974, but disbanded on 30 September 1974 to convert to Phantom FGR.2s, reforming at Coningsby on 1 October 1974 before moving to Leuchars on 3 November 1975.

With the final departure of Lightnings of 29 Squadron at the end of 1974, Wattisham was prepared for Phantom FGR.2s and the first to arrive were with 23 Squadron on 25 February 1976. They were joined by the return of 56 Squadron on 9 July 1976, but 23 Squadron disbanded on 30 March 1983, reforming the same day with Phantom FGR.2s at Port Stanley, for air defence of the Falklands, being based at Mount Pleasant. The departure of 23 Squadron left a gap in Britain's air defence and 15 ex-US Navy surplus F-4Js were acquired, powered by the original J79-GE-10 turbofans, these aircraft equipping 74 Squadron at Wattisham which re-formed in October 1984, and shared the QRA South with 56 Squadron. To protect the Phantoms against a conventional attack, a number of HAS were constructed on the western and southern boundaries of the airfield, but with the end of the Cold War 56 Squadron disbanded on 30 June 1992 and was re-formed at Coningsby as a shadow squadron, responsible for Tornado F.3 training. No.74 Squadron continued air defence operations from the HAS area on the western side of the airfield until its farewell flypast on 1 October 1992 with fixed wing operations at RAF Wattisham ending on 31 October, leaving a number of redundant Phantoms being scrapped on site.

In July 1993 the station was handed over to 3 Regiment, Army Air Corps, to be joined by 4 Regiment, the station being the home for Army Lynx helicopters to be followed by the Army attack AH.1 Apache Longbow helicopters, which are the current residents on this well equipped and modern air base. The QRA shed, which was located on the SE corner of the airfield has been dismantled and re-erected at Bruntingthorpe to house two "live" Lightning F.6s, preserved in running order at the airfield.

Above: Lightning F.1s were with 56 Squadron at Wattisham from December 1960 until April 1965 when the earlier versions were replaced with Lightning F.3s. Heading the slightly uneven flight line here is Lightning F.1 XM174 D:56 Squadron with the Wattisham No 1 'C' Type hangar in the background. *J. J. Halley*

Below: Before the advent of HASs, dispersed aircraft were protected by concrete walls, between which the aircraft were parked. This example at Wattisham is protecting retired Canberra B(I)8 XM275 which had served with 16 Squadron in Germany. *Author*

Chapter 7

RAF Strategic Bomber Stations

It was somewhat surprising that the British Government decided to order three different strategic bombers, known as V-Bombers, for the RAF, although the Valiant was available earlier, at lower technical risk, to fill an urgent gap in strategic bomber capabilities. The Valiant was the insurance against development difficulties with the delta-wing Avro Vulcan, and even more advanced crescent-wing Handley Page Victor. All three were designed to carry nuclear weapons at high altitude against targets in Eastern Europe and the Soviet Union, were unarmed and carried a crew of five. Following the Gary Powers U-2 incident, it was realised that not only could the strategic bombers be detected well in advance by surveillance radars, but the Soviet Union had the capability of sending missiles up high enough to intercept the bombers. As a result, the whole operation changed to low-level penetration to avoid detection by the enemy radars until it was hopefully too late. In due course it was possible for them to carry rocket-powered stand-off missiles, which gave the bomber a better chance of escape from the resulting nuclear explosion, as well as limiting its exposure to enemy defences. The Valiant was the first to suffer from metal fatigue due to the extra stresses

Above: The Vickers Valiant Prototype WB210 powered by four Rolls-Royce Avon RA.3 turbojets of 6,500lb thrust each, made its first flight from the Vickers airfield at Wisley on 18 May 1951. An initial production contract for 25 Valiants was placed on 12 April 1951, and the type entered service with 138 Squadron at Gaydon from February 1955. *Charles E. Brown*

at low altitude, while the stronger Vulcan became the major RAF strategic bomber, supported by the Victor initially in the bomber role, but later as a flight refuelling tanker, until its structural life was also expired.

The airfields used by the strategic bombers were mostly the well-established bomber stations constructed during the RAF Expansion Scheme of the 1930s with modernisation to support the new bombers, which demonstrates the foresight of the original layout and construction, that they are in many cases still serving the RAF today.

Left: The Handley Page Victor prototype WB771 powered by Armstrong Siddeley Sapphire turbojets with some 8,300lb thrust, made its maiden flight from the A&AEE at Boscombe Down on 24 December 1952. An initial production order for 25 Victors was placed in June 1952 and the type entered service with 10 Squadron at Cottesmore on 15 April 1958. *Charles E. Brown*

Above: Coningsby-based 12 Squadron was re-formed on 1 July 1962 with Vulcan B.2s and moved to Cottesmore on 17 November 1964 where it was disbanded on 31 December 1967. Vulcan B.2 XM598 was being prepared for another sortie at Coningsby during 1963, and retains the white overall reflective finish and the squadron crest on the fuselage side.
RAF Museum P013789

Coningsby, Lincs;
TF225565, S of Coningsby village by A153

Ever since it opened as a bomber base in November 1940, Coningsby has been one of the major strategic RAF stations, as home to strike and air defence operations which continue today. Following the disbandment of the nuclear-capable Canberra B.6 squadrons in July 1961, Coningsby was prepared for a three-squadron wing of Vulcan B.2 V-Bombers. The first to reform was 9 Squadron on 1 March 1962, followed by 12 Squadron on 1 July, and the Wing was completed on 1 December when 35 Squadron also re-formed. The stay of the Vulcans was relatively brief, as they departed in November 1964 for Cottesmore, and Coningsby changed role from strategic bomber to air defence with Phantom FGR.2s.

Left: Coningsby opened in November 1940 as a 5 Group Bomber Command station with two 'J' Type hangars as part of a technical site on the northern boundary. It closed in mid-1942 for the construction of three runways, a B.1 and three T.2 hangars, reopening in August 1943. Bomber operations commenced with 617 Squadron in September 1943, and continued with various squadrons until the end of World War 2 with the last operation in April 1945. The Lancasters were replaced by Lincoln B.2s from July 1946, which were then replaced by Washingtons from October 1950. The Washingtons were replaced by Canberras from March 1953 until early 1954 when the station was closed again for improvements including extending the main runway to 9,000 feet and closing the north-south runway. Runway 12 – 30 as shown in the May 1964 airfield chart was retained for a while, but later disused. Canberras returned in November 1956, remaining until July 1961. The first Vulcans arrived in March 1962, and remained until November 1964, when the station was put under C&M in preparation for the operation of air defence Phantoms. *Jeppesen*

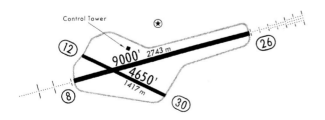

Cottesmore, Rutland;
SK905155, by Cottesmore village W of A1 and N of B668.

Following the departure of the Canberras on 20 February 1955, Cottesmore was extensively developed as a V-Bomber base over a period of three years, including a new 9,000ft main runway, new taxiways and four H-pattern hardstandings, instead of the traditional ORPs, to allow rapid access to the runway. New specialist workshop buildings were constructed to accommodate the new electronic, radar and radio systems servicing, together with new crew briefing facilities and a secure nuclear weapons store on the northwest side of the airfield. The domestic site was increased in size to accommodate the additional staff, while the messes were increased in size and modernised, ready for reopening in March 1958 under the command of World War 2 fighter ace Group Captain 'Johnnie' Johnson. The first arrivals on 9 April were three Victor B.1s for 10 Squadron, which re-formed on 15 April as the first RAF Victor unit, followed by XV Squadron which re-formed on 1 September 1958.

As early as June 1945, Handley Page began studies for swept-wing jet-bomber designs, resulting in the H.P.80 which was submitted to the Air Ministry in March 1946. This stimulated the Air Staff and on 7 November 1946 an Operational Requirement was issued, later confirmed in Spec. B.35/46 for a four-engine jet aircraft capable of carrying a 10,000lb nuclear weapon over a range of 3,500 miles with an operational ceiling of 45,000 feet. The performance parameters were continually updated, resulting in a maximum speed of 525 knots, a maximum range of 5,000 miles and an operational ceiling of 50,000 feet. Power was from what was to become Armstrong Siddeley Sapphire turbo jets, and the fuselage diameter of 9ft 10in was to provide the widest bomb bay of all the V-Bombers, giving a greater load-carrying capacity. The wing was a crescent-shape profile with the inboard section swept back at 48.5 degrees, the centre section swept at 37.5 degrees and the outboard thinner section at 26.75 degrees. The five-man crew consisted of two pilots, a navigator/plotter, a navigator/radar operator for bomb aiming and a radio/electronic countermeasures operator.

It had been originally intended to accommodate the crew in a pressurised escape module, but this was later abandoned, leaving just the pilots with ejection seats, and the rear crew having to abandon by the side door in an emergency. With the final approval of the H.P.80 design by the MOS, two prototypes were ordered on 28 April 1948, and because the Radlett runway needed extending, the first prototype WB771 was dismantled and taken by road to Boscombe Down where it made its short maiden flight on 24 December 1952. Tragically, this prototype crashed at Cranfield during position error tests, when flutter caused the tail to break off. The second prototype was still being built at Park Street, and with pressure brought to bear for an urgent completion it made its maiden flight from the extended Radlett runway on 11 September 1954, just in time to appear at the Farnborough air show that afternoon. With the initial manufacturers' trials completed, WB775 was delivered to the A&AEE at Boscombe Down on 14 March 1955 for service trials.

During the summer of 1952, the final production configuration of the Victor B.1 was established with the first batch of 25 aircraft ordered in June at a unit cost of £450,000. This was to the same specification as the Avro Vulcan, the intention being to order 25 of each, and decide from their performance in service which one would be ordered in quantity. The first Victors were to have a normal all-up weight of 160,000lb, but be capable of an increase to 190,000lb. The weapons load would be a 10,000lb 'Blue Danube' nuclear weapon with an aircraft range of 4,850 nm at 48,000 feet and a cruising speed of 500 kts. The Victor could also carry conventional weapons, examples being four 10,000lb or 48 1,000lb bombs. The first production Victor B.1 XA917 made its maiden flight from Radlett on 1 February 1956, joining the flight development together with later aircraft, and as the production rate built up, so the first aircraft were delivered to 232 OCU at Gaydon, and with the crews fully trained to 10 Squadron at Cottesmore.

The strength of each squadron was normally eight Victors, with four on standby availability, two on base servicing

Left: No XV Squadron re-formed at Cottesmore with Victor B.1s on 1 September 1958 and disbanded on 1 October 1964. Eight Victor B.1s of XV and 10 Squadrons were lined up at Cottesmore with XH648 of XV Squadron nearest. *IWM HU.81577*

Above: With the departure of the second line aircraft, Cottesmore prepared for training crews from Britain, Germany and Italy on the Tornado GR.1 strike aircraft, the unit being known as the TTTE. British marked Tornado GR.1 ZA362 B-09:TTTE taxiing at Cottesmore for its next training sortie. *Author*

amalgamated into the new Strike Command with the bombers part of No.1 Group. By 31 January 1969, the Vulcans of 9 and 35 Squadrons had departed for Akrotiri in Cyprus, with QRAs ceasing at Cottesmore.

With the departure of the V-Bombers, three weeks later Cottesmore was allocated to 90 Signals Group and in April 1969 98, 115 and 360 Squadrons began to arrive, with Canberra E.15s for airfield radar checking, Argosies for navigation aid and approach system checking and the joint RAF/RN flying Canberra T.17s in the electronic warfare role. On 19 May 1969, 231 OCU arrived from Bassingbourn to train all Canberra crews. The 1975 Defence Review resulted in the closure of twelve stations, and major changes for Cottesmore. No 360 Squadron moved to Wyton in August and a reduced 231 OCU departed for Marham in early February 1976. The Argosies of 115 Squadron also moved to Brize Norton in February and finally 98 Squadron disbanded on 27 February 1976, passing its Canberras on to 100 Squadron at Marham. Cottesmore was then placed on C&M from 31 March 1976 until 31 March 1978 when the station was prepared for the Tri-National Tornado Training Establishment (TTTE), where all British, German and Italian crews were trained in the operation of the Tornado with aircraft from all three countries on the strength. With the majority of initial crews trained for the three air forces, RAF crew training became the responsibility of 15 (Reserve Squadron) and the TTTE closed on 31 March 1999.

Cottesmore returned to frontline status on 1 April 2000, when Joint Force Harrier (JFH) was created; the Harrier GR.7s with 4 Squadron had arrived on 13 April 1999 and 3 Squadron on 11 May 1999, both from Laarbruch in Germany, to be joined by 1 Squadron from nearby Wittering on 28 July 2000. With the premature withdrawal of the Harrier FA.2s from RN fleet defence in April 2004, 3 Squadron was disbanded to become the first Typhoon squadron at Coningsby on 31 March 2006, and in its place 800 NAS was commissioned with Harrier GR.9s on the same date. The plan was for 801 NAS to join the JFH, as the Naval Strike Wing, but due to economies, 801 was never commissioned. Cottesmore Harriers played a major part in supporting Allied ground forces in Afghanistan until replaced by Tornadoes in 2010. Cottesmore was the location of the BAE Systems modification line for all Harrier GR.7s to be modified

and two more undergoing major servicing and update modifications elsewhere. To respond to a warning of a nuclear attack, the Victor crews trained for fighter-like scrambles. To achieve getting four Victors in the air in less than four minutes, all four engines could be started simultaneously, in the hope that they would be ready to retaliate with nuclear weapons before the station was destroyed.

With the development of the more effective Victor B.2, C Flight of 232 OCU moved in from Gaydon in September with the first aircraft arriving on 2 November 1961 to undertake intensive service trials with the new aircraft. With three Victor B.2s allocated, the unit also trained staff instructors, moving to Wittering in February 1964, which started the wind-down of the Victor units at Cottesmore. On 1 March 1954, 10 Squadron disbanded, followed by 15 Squadron on 31 October, making room for the Vulcan B.2s from Coningsby, which arrived on 2 November. The new Cottesmore Wing in 1 Group Bomber Command with 9, 12 and 35 Squadrons had both nuclear and conventional roles. The crews were tasked with operational training, stand-by against nuclear attack and also wide-ranging overseas deployments, an example of the latter in 1965, being based in Gan, Butterworth in Malaysia and Takoradi during the Indonesian confrontation.

With the transfer of the nuclear deterrent to the more mobile Royal Navy submarines due to the increasing vulnerability of the easily targeted V-Bomber airfields, reductions in the V-Force commenced. By this time the flight profiles had changed to low-level penetration armed mainly with special conventional weapons and 12 Squadron disbanded on 31 December 1967. On 30 April 1968, in a major reorganisation of the RAF, the traditional Bomber, Fighter and Coastal Commands were

Right: With the departure of 3 Squadron to Coningsby in March 2006, 800 NAS was able to use their Harriers and formally join JFH. Harrier GR.9 ZD329 10:800 NAS was being pulled off the Harrier GR.9 conversion line at Cottesmore ready for its first check flight. *Author*

to GR.9 standard, which involved a major strip down and update. With the Government Strategic Defence and Security Review in October 2010, it was announced that amongst the defence cuts, HMS *Ark Royal* would be decommissioned and the Harriers would be withdrawn from service on 15 December, with 1 Squadron disbanding on 31 January 2011, resulting in the closure of Cottesmore to flying on 31 March and the station being reduced to C&M by 1 November.

Finningley, S Yorks;
SK665980, 7 miles SE of Doncaster, by A614.

Planned as part of the RAF Expansion Scheme of the 1930s, the site was first purchased in September 1935 for an airfield on farmland west of the village of Finningley, and construction commenced as soon as the harvest could be brought in. The technical site was located on the NE corner of the airfield and included five 'C' Type hangars, with four in a semicircle and the fifth located behind the southern end. The remainder of the standard-pattern technical site was located behind the hangars, with the station accommodation beyond. The first RAF personnel moved into the site on 30 July 1936 and the station officially opened on 3 September with the first aircraft, Heyfords of 102 Squadron, having arrived on 31 August. Finningley became a bomber station within 3 Group Bomber Command and a number of squadrons formed up at the station in preparation for World War 2. With the outbreak of war in September 1939, Finningley was tasked with operational training of bomber crews until 8 September 1940, when Hampdens of 106 Squadron started mine-laying operations until February 1941 when it moved to Coningsby, and 25 OTU was formed on 1 March to continue the bomber aircrew training task, which continued until the end of the war.

With the coming of peace, 616 (South Yorkshire) Squadron RAuxAF re-formed at its traditional home base on 6 June 1946 equipped with Mosquito night fighters, having been the first RAF squadron to operate Meteor jet fighters in World War 2. Meanwhile the training role continued at Finningley, with a transfer to Training Command on 1 May 1947; this continued until 21 April 1954, when all regular RAF flying ceased, except for 616 Squadron, and the station was placed under C&M on 1 June. No 616 Squadron converted to Meteors, continuing to operate from the station at weekends until it moved to Worksop in May 1955.

Finningley was then selected to become a V-Bomber base and became part of 1 Group Bomber Command. As part of the overall improvements a new 8,993ft-long main runway was constructed and the other two runways were disused. Finningley reopened in May 1957 and on 1 October the first Vulcan B.1 arrived, leading to the reforming of 101 Squadron on 15 October. As more Vulcans were delivered, the crews were trained in the operation of the new V-Bomber, their duties including a number of overseas navigation exercises, and flight demonstrations to show some of the capabilities of this new delta-wing strategic bomber. A second V-Bomber unit arrived on 16 December 1958, when 18 Squadron re-formed with special Valiants equipped with radio and radar jamming equipment instead of bombs, for electronic counter-measures (ECM) duties. In February 1960 the Bomber Command Development Unit (BCDU) arrived from Wittering equipped with a small number of Canberras, Valiants and Vulcans to undertake trials with new systems being developed for the V-Force. On 17 June 101 Squadron moved to Waddington, and in its place 230 OCU arrived two days later from Waddington, Finningley returning to its operational training role. With the operational V-Bombers carrying their own ECM systems, 18 Squadron was disbanded on 31 March 1963, leaving 230 OCU

Right: Vulcan B.2s of 230 OCU-based at Finningley, but on the Waddington ORP with the Lancaster overflying on the occasion of the formation of Strike Command on 24 April 1968. *IWM RAF T.8296*

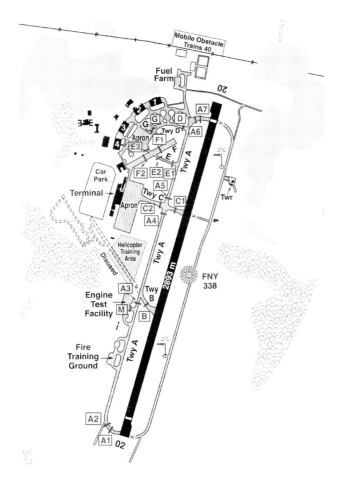

Above: Finningley first opened in September 1936 as part of the RAF Expansion Scheme with five 'C' Type hangars in the northwest-located technical site, and during World War 2 was used for bomber crew training. It was placed under C&M in April 1954 and allocated to 1 Group Bomber Command in May 1955 to be developed for V-Bombers. The main runway was extended to 8,993 feet and the other two disused, reopening in May 1957 ready for 101 Squadron to reform with Vulcans in October. In June 1960 230 OCU arrived for Vulcan air and ground crew training and the operational V-Bombers departed. Finningley was allocated to C&M in December 1969 and was taken over by Training Command in May 1970. The station finally closed in November 1996 and in April 2005 it reopened as Doncaster Sheffield (Robin Hood) Airport. *Pooley's Flight Guides*

as the main resident, accompanied by the BCDU with a few aircraft. In June 1963 the Bomber Command Vulcan Servicing School arrived as a non-flying unit to train the engineering support personnel to provide maintenance support for the Vulcan fleet. In July 1965 232 (Victor) OCU disbanded at Gaydon, and ground training on the type was moved to Finningley.

On 1 November 1965, Finningley became responsible for the Bomber Command detachments at Goose Bay in Labrador and Offutt in Nebraska where V-Bomber crews under training operated long-range flights into unfamiliar territory, and low-level flights could be flown over the desolate uninhabited territories. On 30 April 1968, Bomber Command became part of Strike Command and on 31 December the BCDU was disbanded. In August 1968 the Vulcans of No 1 (Bomber) Group Standardisation Unit (GSU) had moved in from Waddington, and they were joined by the Victor section from Marham, the duties of the unit being to maintain standards of operation of all the Vulcan and Victor squadron aircrews, whether in the bombing or flight refuelling tanker role. In September 1969 1 GSU moved out and in December 230 OCU moved to Scampton, the station being placed under C&M, its bombing operational role completed. Finningley was then transferred to Training Command and in May 1970 6 FTS re-formed with Varsity navigational trainers. In October 1973, Finningley became responsible for all non-pilot aircrew training for the RAF when the Air Electronics and Air Engineers School was transferred to Finningley. The Varsities were replaced by Dominie navigation trainers in April 1976. On 29 July 1977, Finningley was the venue for HRH The Queen's Silver Jubilee Review of the RAF when all flying squadrons in the RAF were represented. On 23 April 1979, the training role increased further with the arrival of the Jetstreams of the Multi-Engine Training Squadron (METS) to train student pilots to operate multi-engine aircraft, making Finningley one of the busiest RAF stations.

Despite all this flying activity, which included the training of fast jet navigators in Jet Provost T.5s, which were replaced by Hawks and finally Tucanos, in 1994 the MOD announced under the Front Line First defence cuts that Finningly would be closed, which finally happened in November 1996. Three years later, the site was acquired by Peel Holdings with plans to open it as a commercial airport, and on 28 April 2005, it reopened as Robin Hood Airport, serving Doncaster and the surrounding area.

Gaydon Warks;
SP355555, 8 miles S of Warwick, by A41.

Gaydon's main claim to fame was that it was selected as the first V-Bomber base, but it had a rather pedestrian beginning. The station opened on 13 June 1942 with three concrete runways and basic accommodation including T.2 and B.1 hangars and 27 heavy bomber-type hardstandings for the training of bomber crews in Wellingtons. This continued until 1 July 1945, after which the station was transferred to Training Command until closed to flying on 15 August 1946, and being placed under C&M from 28 August. In 1953 development of Gaydon commenced in preparation for the arrival of the Valiant V-Bombers. Amongst the changes was a completely new 9,000ft runway on the southern boundary of the original airfield with three groups of four dispersals leading to the main runway for quick response departures. A 'Gaydon' Type hangar

Above: The Valiant B.1s were introduced into RAF service by 138 Squadron at Gaydon which re-formed on 1 January 1955. With its compliment of ten Valiants, the squadron was on parade at Gaydon on 29 June 1955. It moved to Wittering on 6 July 1955 to make room for the Valiants of 232 OCU, which was responsible for the training of air and ground crews. *RAF Museum P004471*

Below: Initial operations by 138 Squadron were from Gaydon which flew Valiant B.1s from February 1955 until March 1962. The squadron moved to Wittering on 6 July 1955, and this Valiant in the original silver finish has just finished a sortie and the crew, headed by Wing Commander Oakley (second from the right) are returning for the debrief. The wing commander's pennant is on the side of the fuselage close to the door. *RAF Museum P004476*

was constructed, similar to the 'J' Type to accommodate the Valiants and a new operations block and electronics building were constructed as well as a secure weapons store. With the increased size of the station more messes and domestic accommodation was constructed.

Gaydon was reopened on 1 March 1954 and on 1 January 1955 138 Squadron re-formed as not only the first Valiant unit, but the first in the V-Force. After working up to operational status, 138 Squadron moved to Wittering on 6 July 1955, and meanwhile 232 OCU formed on 21 February to train further Valiant crews. This included 543 Squadron which re-formed at Gaydon on 1 July 1955 to operate Valiants in the strategic reconnaissance role, moving to its permanent base at Wyton on 18 November 1955.

Six British aircraft companies submitted tenders for Specification B.35/46 in 1947 for a four-jet-powered medium bomber. The aircraft selected were the A. V. Roe Vulcan and Handley Page Victor, but both were recognised as being very advanced designs and may not only carry some risks, but would also leave an unacceptable strategic nuclear bomber gap with the RAF. It was therefore decided in September 1957 to order the development of a third type to a less advanced design with a relaxation in overall performance, and Vickers was selected in January 1948 to proceed with the construction of two prototypes and Specification B.9/48 was issued on 19 July to cover the new aircraft. The capability was reduced to carrying up to one 10,000lb nuclear weapon over ranges of 1,500 nm at a cruising speed of 535 mph/862 km/h at a height of up to 45,000 feet. The crew of five consisted of two pilots, two navigators and protective systems operator. Power came from four Rolls-Royce Avon turbojets, the production versions being fitted with Avon 204s developing 10,050lb thrust. Weapons delivery was by an H2S radar blind bombing aid which had been developed during World War 2, and the aircraft was unarmed.

To avoid delays with entry into service of the Valiant, instead of waiting until flight trials had proved the concept, a production contract was placed in February 1951 for an initial 25, followed by 24 more B.1s in October, and 17 of the proposed B.Mk.2. The first prototype made its maiden flight from Wisley on 18 May 1951, seven months ahead of programme, this aircraft being intended for aerodynamic and performance development, and was therefore not fully equipped with its operational systems. On 11 April 1952, the more fully equipped second prototype flew three months ahead of schedule, although the first prototype had been lost on 12 January, bringing pressure on the flight development programme.

In April 1953, approval was given for another 56 Valiant B.Mk.1s, but in June this was altered to 53 B.Mk.1s and nine -reconnaissance B(PR).Mk.1s with sensor fit in the bomb-bay instead of weapons. Following further adjustments, a total of 104 production B.Mk.1s were built, and the first production Valiant made its maiden flight on 21 December 1953, nine months later than planned due to changes made on the production line as a result of flight trials. By June 1955, 16 Valiants were engaged on various test programmes, and limited CA Release was achieved in January 1955. The last production Valiant was flown from the short runway at Brooklands, where the production line was located, on 27 August 1957 to the nearby test airfield at Wisley, the build rate having reached one aircraft per week.

With the formation of 138 Squadron at Gaydon, deliveries continued with seven Valiants delivered by early June 1955, and in addition to the formation of 232 OCU, the Bomber Command Development Unit (BCDU) was also based at Gaydon. In September two of 138 Squadron's Valiants took part in the type's proving trials with flights to Singapore, Australia and New Zealand without any serious problems. In the spring of 1956, 138 Squadron flew four aircraft on 1,000 hours intensive flying trials as the final test to prove the bomber's capabilities.

Once the majority of the Valiant crews had been trained, 232 OCU took over the responsibility for the training of the crews for the Victor, XA931, the first of seven aircraft arriving at Gaydon on 28 November 1957. In addition to the training tasks, 232 OCU also carried out a period of intensive service trials to prove the production standard aircraft, the OCU continued the training of all Victor B.1 crews, as well as converting those from the Valiants squadrons. This continued until 30 June 1965, when 232 OCU disbanded. The station was then taken over by Training Command and on 1 September 1965 the Varsities of 2 ANS arrived until they departed to Finningley on 1 May 1970. Various RAF units passed through Gaydon until 1 April 1972 when it was put under C&M and finally closed on 31 October 1974. The site was acquired by British Leyland and developed for vehicle testing, which still continues under the ownership of Jaguar Land Rover.

Honington, Suffolk;
7 miles NE of Bury St Edmunds.

With the preparations completed to accept the first V-Bombers, including a secure weapons store on the northern boundary, on 1 November 1956 7 Squadron re-formed at Honington and the following month, the first Valiant B(PR).1s arrived, starting a major build up of strategic nuclear-capable bombers at the station. The new main runway was 9,012ft long with over-runs taking its total length to 9,308 feet, connected by a concrete perimeter track with bomber dispersals around the northern side of the airfield. No. 7 Squadron was joined by the Valiant B(K).1s, B(PR).1s and B(PR)K.1s of 90 Squadron when it re-formed on 1 January 1957, by which time 7 Squadron was operating the air-to-air flight refuelling Valiant B(K).1. No. 199 Squadron re-formed at Honington on 1 October 1957 equipped with Valiants and Canberras fitted with electronic warfare systems. However, 199 Squadron disbanded on 15 December 1958, its role being taken over the next day by a re-formed 18 Squadron at Finningley.

Further expansion of the V-Force at Honington continued with 57 Squadron reforming on 1 January 1959 and 55 Squadron reforming on 1 September 1960, both with the Victor B.1a V-Bombers. With the preparations for the V-Force at Wittering completed, 7 Squadron moved there on 1 September 1960 and 90 Squadron disbanded at Honington on 1 May 1965. The Honington V-Force squadrons maintained a high alert status as a deterrent against Soviet nuclear attack, with many exercises held to keep the crews at readiness, the aim being to get at least four aircraft airborne within

Left: Honington from the air c.1960 showing the traditional RAF Expansion layout including the gentle semi-circle of five 'C' Type hangars behind the vast ASP. Seven distinctive 'H' block barracks can be seen and to the left of them is the main entrance with the guardroom and station HQ. The six aircraft on the ASP appear to be Phantoms with a Nimrod at the rear. *RAF Museum P017590*

Left: Although the Victor was designed to deliver nuclear weapons, it could also carry a substantial load of conventional bombs, demonstrated by 57 Squadron Victor B.1 XH648. No 57 Squadron was re-formed at Honington on 1 November 1959 with Victor B.1s which it operated until June 1966. The squadron moved to Marham on 1 December 1965. *IWM HU.81578*

less than four minutes. When the nuclear deterrent was taken over by the more mobile RN nuclear submarines both the Victor units were transferred to the in-flight refuelling role, replacing the Valiants which were suffering from structural fatigue. The centre for flight refuelling was at Marham, and 55 Squadron departed for there on 24 May 1965, followed by 57 Squadron on 1 December. Honington was then put under C&M until 1969, and in 1967 it was earmarked as the base of the RAF F-111K swing-wing bombers, but this programme was cancelled on 16 January 1968. Instead of the American aircraft, which was having development difficulties at the time, the Blackburn Buccaneer S.2 was chosen for the RAF as a low-level strike aircraft in the maritime role. Honington re-opened on 1 January 1969, and on 1 October 12 Squadron re-formed with its first four Buccaneers.

The Buccaneer S.1 had originally been designed for the FAA to operate from RN carriers, and with two crew was powered by the inadequate Gyron Junior turbofans. No. 12 Squadron was equipped with the more effective ex FAA Buccaneer S.2s and assigned to Supreme Allied Commander Atlantic (SACLANT) under the operational control of 1 (Bomber) Group. Exactly one year later XV Squadron became the second RAF Buccaneer unit also at Honington, equipped with the new build S.Mk.2B aircraft. After working up with the new aircraft, XV Squadron moved to RAF Laarbruch in Germany on 11 January 1971, where it was assigned to Supreme Allied Commander Europe (SACEUR) as part of the 2nd Allied Tactical Air Force. In the place of XV Squadron 237 OCU was formed at Honington on 1 March 1971 equipped with ex FAA Buccaneer S.2s to be responsible for the training of crews to operate the new strike aircraft, both for the RAF and remaining Buccaneers in the FAA, the instructors coming from both the RAF and FAA.

With the cancellation of the TSR.2 in April 1965, followed by the cancellation of the Royal Navy's new aircraft carrier and the cancellation of the F 111K contract, 26 Buccaneer S.2Bs were ordered for the strike role with the RAF. This was soon followed up by an order for a further 17 aircraft, and with the loss of RN capability of operating carrier born fixed wing aircraft, 64 of the FAA Buccaneers were to be transferred to the RAF. With the withdrawal of the carriers, the RAF would gradually take over the maritime strike/attack operations from the RN.

Right: No 13 Squadron re-formed at Honington with Tornado GR.1s on 1 January 1990 and moved to Marham in February 1994. *BAe*

Left: No IX Squadron re-formed at Honington on 1 June 1982 with Tornado GR.1s and moved to Germany in 1986. Two Tornado GR.1s from Honington-based IX Squadron are carrying four 1,000lb bombs under the fuselage, two underwing fuel tanks and a pair of ECM pods under the outer wing. *BAe*

To reduce costs and speed delivery of the new Buccaneer S.2Bs to the RAF, many of the naval features, such as folding wings and arrester hooks were retained, but the wings were modified for the carriage of four Martel anti-ship missiles, and the undercarriage was also strengthened to cope with the increased all-up weight. The range was extended significantly by the development of conformal 425-gallon fuel tank in a fairing over the outer skin of the rotatable bomb-bay door with insignificant increase in drag. While production of the Buccaneers was in the old Blackburn factory on the banks of the Humber Estuary, the airfield was too short for the operation of this sophisticated strike aircraft, so when complete, the aircraft were towed by road with the wings folded to the nearby ex-RAF World War 2 airfield of Holme-on-Spalding Moore where they were flight tested. The first Buccaneer for the RAF made its maiden flight on 11 February 1969, while some RAF aircrew had been seconded to FAA squadrons since 1965, the numbers increasing significantly by 1969 when recruitment numbers for the FAA reduced with planned phase-out of the carriers. Many of the RAF crews were required to operate the Buccaneers with the FAA for two to three years until the withdrawal of the carriers, resulting in a pressing need to train crews for the newly forming squadrons which were also starting in 1969.

To achieve the earliest possible service entry date for the RAF Buccaneers, due to the capability gap by the previous cancellations, the crews for the first two RAF squadrons were trained by 736 Squadron FAA, and the increased training task had to be resourced, by adding experienced RAF aircrew to the training unit. For the training of the RAF aircrew, a batch of six Buccaneer S.1s were brought out of storage and refurbished ready for the first long course to start at Lossiemouth on 19 May 1969, with the final three sorties on Buccaneer S.2s for familiarisation on the operational version. Over a successful two-year period the crews for the first two RAF squadrons were trained before the formation of 237 OCU. Following a catastrophic turbine-fatigue failure in a Gyron Junior engine on 8 December 1970, the Buccaneer S.1s were withdrawn, leaving all training on the more representative S.2 versions.

On 1 July 1974, 208 Squadron formed at Honington with the Buccaneer S.2A to operate in the overland role for SACEUR covering operations in the AFNORTH region, to compliment the Buccaneer squadrons in Germany. No. 208 Squadron had to be able to operate over much greater ranges, making regular use of air-to-air refuelling as well as using airfields in Denmark to cover operations across the Baltic. It was also considered likely that any attack would be from the far north of Norway, requiring support of Allied forces, which presented a very different challenge due to the different terrain and weather with major variations in daylight throughout the year. Distances were vast, with the introduction of the bomb-bay fuel tank a bonus, and the range was further increased by the carriage of underwing fuel tanks. Many of the flights required landing in Norway to refuel both out and back from sorties. The threat differed from the Central Europe region, in that there were no massed air defences, and approaches to targets could make use of terrain masking, allowing a greater degree of surprise attack. However, the individual potential targets were heavily defended, and the element of surprise would be the only chance of survival of the Buccaneers until the introduction of stand-off weapons. The main enemy targets were airfields, radar systems and SAM sites, and armed with BL 755 cluster bombs, the squadron would undertake interdiction sorties against advancing armour formations or supply convoys, as well as countering the difficult task of amphibious landings.

Buccaneers regularly participated with success in the Red Flag exercises at Nellis in the USA, but on 7 February 1980, XV345 suffered a catastrophic failure of the starboard wing with the death of the crew. The Buccaneers were grounded for inspection to the front spar of the inboard wings and the least affected were modified, allowing some 60 aircraft to return to service. No. 216 Squadron had been re-formed in the overland role on 1 July 1979, but with a shortage of aircraft, it was disbanded in August 1980.

With the cancellation of the TSR.2 and then the F-111K for the RAF, followed by the collapse of the Anglo-French AFVG, Britain was in urgent need of a strike aircraft to replace the aged Canberras. Therefore in mid-December 1968 Britain's BAC, Germany's MBB and Fiat of Italy formed a joint company and created a tri-national strike aircraft requirement. By combining the features of the British and German designs, the aircraft became what was to be known as the Multi-Role Combat Aircraft (MRCA) with power from two Rolls-Royce RB.199 turbofans. The overall configuration was agreed on 14 March

1969 with the British-proposed two-tandem-seat cockpit. The joint international company became Panavia on 26 March 1969 with three assembly lines in Britain, Germany and Italy for political reasons, and the governments set up a single customer organisation known as NAMMO (NATO MRCA Management Organisation). Both Panavia and NAMMO were accommodated in the same building in Germany. The aircraft was of traditional metal construction without the use of composites in its structure and its prime aerodynamic feature was to have a variable-geometry wing with a maximum leading-edge sweep of 67 degrees. Although originally named Panther in April 1969, the aircraft soon became known as Tornado. The first Tornado prototype was completed at the MBB factory at Ottobrunn and taken to Manching on 12 November 1973 for assembly, and after some delays, was first flown by Paul Millet of BAC, with Nils Meister of MBB in the rear seat on 14 August 1974. The second aircraft, P.02/XX946, was British-assembled and again flown by Paul Millet from Warton on 30 October 1974, with P.03/XX947 following on 5 August 1975. A total of nine prototypes were built, three in Germany, four in Britain and two in Italy, sharing different aspects of the development programme. These were followed by six pre-production aircraft, of which three were assembled in Germany, two in Britain and one in Italy.

Tornado production was in nine batches consisting of two basic versions; the Air Defence Variant (ADV) was purely for the RAF and details of this type are covered under Coningsby where it first entered service with the RAF. The Interdictor Strike (IDS) version, initially known for the RAF as the GR.1 (Ground Attack/Reconnaissance Mk.1) was produced in four role-specific types; overland strike, electronic warfare, reconnaissance and maritime strike. The initial batch included 35 IDS aircraft, including 12 twin-control-column versions, all of which were allocated to the TTTE at Cottesmore for training. The second batch included 55 IDS aircraft, of which 16 were twin-stick, these aircraft mostly going to the TWCU at Honington, as well as 15, 27 and 617 Squadrons. The RAF eventually received a total of 308 Tornados, out of the overall production of 978 aircraft, with the balance going to Germany, Italy and export to Saudi Arabia.

The most important system in the Tornado was the nose-mounted radar which had a ground-mapping capability to assist with navigation, and a terrain-following capability to give an all-weather round-the-clock low-level capability. While the pilot was logically in the front cockpit, the navigator behind him was responsible for what is known as Mission Management, guiding the pilot to and from the target. In the strike role, the RAF GR.1s could carry one WE.177B free-fall nuclear weapon which was finally withdrawn in mid-1998. The aircraft could also carry a range of 500 and 1,000lb free-fall or parachute-retarded bombs, and for greater precision, Paveway II/III guidance units could be fitted. The Tornado GR.1s equipped eleven frontline strike squadrons, of which eight were originally based in Germany, plus two reconnaissance squadrons.

Preparations were made at Honington for the arrival of the Tornados, with the construction of HAS to accommodate a pair of aircraft, the first GR.1 arriving on 29 June 1981 for the RAF-only Tornado Weapons Conversion Unit (TWCU), which formed at Honington on 3 August 1981 with training commencing on 12 January 1982. No. IX Squadron re-formed with Tornados on 1 June 1982, operating from the northern HAS site until it moved to Germany in 1986. To make room for more Tornados, the Buccaneers began to move out to Lossiemouth, with 12 Squadron going in October 1982, 208 Squadron following in the same month, and finally 237 OCU in October 1984. Meanwhile on 1 January 1984, the TWCU became 45 (Reserve) Squadron, and was renumbered XV (Reserve) Squadron on 1 April 1992 before its move to Lossiemouth. On 1 January 1990, 13 Squadron re-formed with Tornado GR.1As with the combined nuclear daylight-attack role and low-level night intelligence-gathering task and the Honington Tornados were deployed to Operation Desert Storm in January 1991. Following its return from the Iraq conflict, 13 Squadron moved to Marham in February 1994, leaving Honington with no regular based aircraft. The station was then taken over as the RAF Regiment HQ Depot, which continues today, the runway being occasionally used by RAF transport aircraft and helicopters.

Lossiemouth, Moray;
NJ210695, 4 miles N of Elgin, S of B9040.

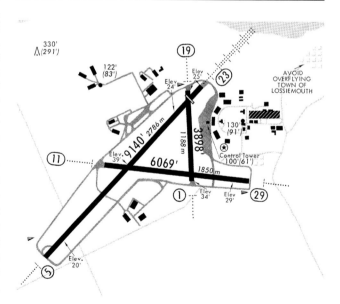

Above: The February 1969 airfield chart shows Lossiemouth with three runways operational, the main one being 9,140 feet long. Runway 19 – 01 was later disused. The main technical site was located on the eastern side of the original airfield with the MU dispersals to the south and northwest. *Jeppesen*

Left: Single-seat Jaguar GR.1s and two-seat Jaguar T.2s of 226 OCU at Lossiemouth from January 1974 were responsible for the training of all RAF Jaguar air and ground crews. MOD

at 5MU Kemble, the selected Shackleton MR.2s were flown to Woodford for conversion to the new role, the first example, WL745 making its first flight on 30 September 1971. The major external change to the aircraft was the installation of a large ventral radome that housed the 1940s-vintage modified AP/ANS radar which was salvaged from the redundant Gannets. The initial role was AEW cover for the RN, later being expanded to provide AEW cover for the whole of the UK defence region. This included the airborne control of air defence aircraft, direction and control of strike/attack aircraft, surface forces observation by Electronic Surveillance Measures (ESM) and a limited capability in the air/sea rescue role. A crew of nine was normally carried, consisting of two pilots, radio navigator, aircraft navigator, engineer and four radar operators. The Shackleton AEW.2s were operated by 8 Squadron, the first of two unconverted MR.2s being delivered for training on 1 November 1971, followed by the first operational AEW.2 on 11 April 1972 while the squadron was still at Kinloss waiting for work to be completed at Lossiemouth. By the end of the year 8 Squadron had eight aircraft on charge, with the final trials aircraft arriving on 17 September 1973, while the squadron was resident at Lossiemouth from 14 August 1973. In the 1981 Government defence cuts, the Shackleton fleet with 8 Squadron was halved to six aircraft, with the remaining aircraft continuing in service for another ten years, their retirement from service being delayed due to the failure of the Nimrod AEW.3 programme, which was cancelled in January 1987. In its place the Boeing Sentry AEW.1 was ordered with the first example arriving at Waddington in early November 1990 and the Shackletons were retired after 19 years interim service, the re-equipment being completed with Sentry AEW.1s by 1 July 1991.

The station was constructed as a permanent RAF station, opening in 1939 as part of Flying Training Command, later to become a bomber aircrew training base. The technical site was located on the east side of the landing field with three 'C' Type hangars in a semicircle and austerity technical and administration buildings due to the start of World War 2. No. 46 MU was opened on 1 April 1940, the first of five 'J' Type hangars being ready for use in June. In addition seven 'L' Type storage hangars were constructed on dispersed sites to the south and west of the airfield, and three concrete runways were laid in 1942 to cope with heavy bomber operations. With the closure of the bomber training operations, the station was taken over by Coastal Command for operational training which continued until August 1946 when the Fleet Air Arm arrived to use Lossiemouth for training and the RAF MU closed in February 1947. The RAF MU facilities were taken over by the FAA as an Aircraft Holding Unit, and the airfield underwent a major expansion programme in 1953, which included the extension of the main runway and east-west runway to allow operations by the Scimitar and Buccaneer service trials prior to the aircraft entering frontline service.

On 29 September 1972, Lossiemouth was returned to the RAF and the station closed for major improvements in preparation for the arrival of the first RAF Jaguars, although the first RAF unit to come to the station was 8 Squadron from nearby Kinloss with Shackleton AEW.2s from 17 August 1973. With the withdrawal of the fixed-wing RN aircraft carriers and the retirement of the FAA Gannet AEW.3s, a gap resulted in Britain's Airborne Early Warning capability before the arrival of the planned Nimrod AEW.3. To fill this gap in 1967 twelve low-flying-hours Shackleton MR.2s were selected for conversion to the AEW.2 as an interim answer to the requirement. Following preparations

From January 1974, what was to become 226 OCU, was responsible for the pilot training on Jaguars, with 54 Squadron forming up in March before moving to Coltishall in August, and followed by 6 Squadron from October to November before it too moved to Coltishall. No. 226 OCU continued with Jaguar pilot training until 1991, when it was renumbered 16(R) Squadron, and was finally disbanded in 2005 as the Jaguar fleet was withdrawn. RAF Buccaneers arrived from Honington from October 1982 until the arrival of 237 OCU in October 1984, the latter disbanding in 1992. With the retirement of the RAF Buccaneers, 12 Squadron re-equipped with Tornado GR.1Bs on 1 October 1993 and moved north in

Left: No 8 Squadron moved from Kinloss to Lossiemouth with Shackleton AEW.3s on 17 August 1973 and this 'interim' type remained in service until replaced by the Sentry AEW.1 from November 1990. Lossiemouth-based Shackleton AEW.2 WL756 with 8 Squadron on patrol off the coast of Britain. *HSA*

January 1994, while the TWCU/XV Squadron arrived from Honington in November 1993, adding to its tasks the full crew training role on the Tornadoes when the TTTE disbanded at Cottesmore in 1999. No. 208 Squadron was the final RAF Buccaneer unit which disbanded in April 1994, to be replaced on 27 April at Lossiemouth by 617 Squadron Tornado GR.1Bs from Marham.

The Maritime Strike Tornados at Lossiemouth took over the role from the Buccaneers when they were retired on 27 March 1994. With a pool of surplus aircraft available following the withdrawal of 15, 16 and 20 Squadrons from Germany as east/west tensions reduced, a batch of aircraft were converted to the GR.1B standard with the new versions supplied to 12 and 617 Squadrons. The GR.1B was capable of carrying up to four Sea Eagle sea-skimming anti-ship missiles. The aircraft could strike over 400 miles from take-off, and the missiles had a stand-off range of over 70 miles. The aircraft were basically standard GR.1s with systems installed to allow the operation of the Sea Eagle weapons. A total of 27 Tornado GR.1s were converted, the first example making its first flight on 18 September 1993 and each squadron at Lossiemouth had twelve aircraft on strength with 15 crews, although some of the aircraft were deployed to various international trouble spots.

The task of the Tornado GR.1Bs was to attack the largest enemy naval assets, which required as many missiles in the air as possible to saturate the enemy defences. To achieve this could require as many as six GR.1Bs operating at low level in widely spaced pairs with the missiles updated on the target position by the aircraft systems. The launch position was designated and once departed from the aircraft, the missile was a fire-and-forget system, the salvo of missiles arriving on target +/-5 seconds apart. The maritime units were not just limited to overwater operations, but retained their interdiction and strike roles against land-based targets.

With the withdrawal of the Tornado force from Germany, 14 Squadron moved from Bruggen to Lossiemouth in January 2001, the airfield therefore becoming a major active RAF station with three front line Tornado squadrons and the training unit. With the Government's SDSR in October 2010, there were suggestions that maybe two squadrons of Tornadoes could be disbanded threatening the closure of Lossiemouth, but this would have brought economic hardship to the area with also the closure of Kinloss. It was therefore decided to possibly close Leuchars, and station two Typhoon squadrons at Lossiemouth, with one squadron of Tornadoes, which will probably be the OCU. In the event, both 13 and 14 Squadrons, the latter at Marham, will cease operations on 1 June 2011, leaving Lossiemouth with the two maritime-strike 12 and 617 Squadrons, and XV(R) Squadron to provide the training, but at a reduced scale.

Marham, Norfolk;
TL730685, 10 miles E of Downham Market, N of A1122.

With the arrival of the first Valiant V-Bombers on 15 March 1956 for 214 Squadron which had re-formed on 21 January, Marham took on a strategic role although the initial build up of the V-Force was slow. On 1 April 207 Squadron re-formed, receiving its first Valiant in July followed by 148 Squadron reforming on 1 July. The first potential action for the Valiants was Operation Musketeer when Egypt attempted to nationalise the Suez Canal, and 16 Valiants from the three squadrons were deployed to Malta. To maintain the nuclear deterrent and raise the airfield standard up to a Class 1 Medium Bomber capability, various improvements needed to be made, including a four-track fast-getaway ORP beside the NE end of the main runway. With the more advanced Vulcan and Victors becoming available for service, the Valiants were transferred from their strategic role to a tactical one of flight refuelling tankers. The application of flight refuelling capability for all V-Bombers had been agreed in 1954, and in December 1957, 214 Squadron was tasked with pioneer trials and development of the tanker capability, but it was not until January 1959 before fuel was transferred between two Valiants in flight. A

Left: No 214 Squadron operated Valiant B(PR)k.1s at Marham from April 1956 until December 1964. It re-formed on 21 January 1956 and disbanded on 1 March 1965 having pioneered RAF air-to-air flight refuelling operations. Valiant B(PR) K.1 WZ390 214 Squadron being towed at Marham by the battery of Bloodhound air defence missiles. *RAF Museum P015122*

total of 16 Valiants were converted, allowing Honington-based 90 Squadron to also change to the flight refuelling role.

On 1 January 1960 207 Squadron changed to SACEUR control in a tactical role, followed by 148 Squadron on 15 July 1961. As a result, one aircraft and crew would be on QRA at 15 minutes readiness continually. On 26 June 1961, 49 Squadron arrived at Marham from Wittering, the 24-strong Valiant force replacing 24 Canberras previously tasked to attack tactical targets. As a result the primary weapon of a single Blue Danube nuclear was replaced by two American 1,900lb Mk.28 nuclear weapons which would normally be released from 40,000 feet. To help to evade the improved Warsaw Pact defences, a hi-lo-hi flight profile was adopted from April 1963, by now carrying the US Mk.43 nuclear weapon.

No 214 Squadron lost its bomber role in April 1962 to undertake air-to-air refuelling (AAR) exclusively, leaving the other three squadrons as the Valiant Tactical Bomber Force (TBF). Then disaster struck when a 543 Squadron PR Valiant was discovered to have a crack in the main wing spar. These aircraft normally operated at higher altitudes, so the TBF Valiants, with their low-level operations, for which they had not been designed, were considered more vulnerable. On 6 August 1964 serious metal fatigue was found in a Marham Valiant, and following detailed inspections of all other Valiants they were found to be damaged, resulting in flying limitations being imposed. Valiants were finally grounded on 26 January 1965, as they were uneconomical to repair, and 214 and 207 Squadrons were disbanded on 28 February, with 49 and 148 Squadrons on 1 May. All of Marham's Valiants, apart from XD818 which was preserved as the sole example to survive, were scrapped on the south-east corner of the airfield.

The loss of the Valiants left a serious gap in AAR capability and a batch of Victor B.1s were converted to the AAR role at Radlett. Many of the Victor Mk.1s were already at Radlett following the disbandment of the Cottesmore Victor Wing. The configuration decided upon was a three-point tanker with a hose and drum unit mounted in the rear of the bomb bay, and two underwing hose and drogue units mounted under the outer wings. The fuselage-located unit would be used for topping up the larger transport and bomber types, while the wing-mounted units could refuel the fighters and tactical strike aircraft. However with the urgency to support the RAF defensive fighter patrols, six B.1As were converted into interim two-point tankers with bomb-bay fuel tanks and FR pods under the wings. The first conversion, XH620, flew on 28 April 1965.

Six Victor B.1s arrived for 55 Squadron on 24 May 1965 for training, and in the following month the first two-point Victor B(K)1A arrived with flight refuelling equipment fitted under the outer wings, the other five arriving over a period of two months. With the arrival of the tankers, 55 Squadron commenced a period of trials with Lightning fighters to establish standard operating procedures and undertook trials to integrate the flight refuelling capabilities within the overall UK Air Defence Region. In June the Tanker Training Flight (TTF) was formed to convert crews to the AAR role and on 1 December 1965, 57 Squadron re-formed with Victor B.1/1As to become the second Victor AAR unit. Meanwhile at Radlett, work started on the modification of the Victor B.1s and improved B.1As into three-point tankers, ten of the former and 14 of the latter being completed, the first production conversion, K.1 XA937, flying on 2 November 1965, going to Boscombe Down for service trials.

The following month, the first three-point tanker was received by 57 Squadron with an additional hose unit fitted in the rear fuselage. On 1 July 1966, 214 Squadron was re-formed with Victor B(K)1 three-point tankers as the third AAR squadron, all the AAR squadrons supporting overseas deployments and the topping-up of Lightnings and Phantoms defending against Soviet raiders in the UK ADR. When 55 Squadron received its three-point tankers, the earlier versions were passed to the TTF which was redesignated 232 OCU on 6 February 1970 with the task of all AAR training which continued until disbandment on 4 April 1986.

The definitive tanker version of the Victor was based on the B.2, whose more powerful engines would allow a greater fuel load to be carried, and improved take-off performance from hot and high airfields. The Victor B.2s were still in operation at

Left: A line-up of 214 Squadron Victor BK.1As at Marham with XH621 nearest on 26 August 1976. *Author*

Wittering until the end of 1968, but XM715 was under repair at Radlett and was used for trial installation for the three-point tanking role. Additional fuel could be carried in fixed underwing slipper tanks mounted at about half span, and all the offensive systems were removed. A total of 24 aircraft were to be converted, with design work undertaken at Radlett, where the aircraft of the disbanded Wittering Wing had been stored since the beginning of 1969. Handley Page had tried to remain independent of the Government-inspired aerospace mergers, but with no guarantee of further work, went into liquidation on 27 February 1970. The contract for the conversions was then awarded to Hawker Siddeley Aviation at the old Avro factory at Woodford, and the stored B.2s were prepared for ferry flights and flown to Woodford between July and August 1970, the first Victor K.2, XL231 making its maiden flight as a K.2 on 1 March 1972.

The replacement of the Victor B(K).1s commenced with the far more capable Victor B(K).2s with the arrival at Marham of XL233 for 232 OCU on 7 May 1974. As the replacements came off the conversion line at Woodford, 55 Squadron was the first to convert to K.2s on 1 July 1975, to be followed by 57 Squadron a year later. Because the relatively small number of Victor K.2 conversions, 214 Squadron retained the Victor K.1s until the other two squadrons were fully equipped and it was disbanded on 28 January 1977 after being responsible for developing the concept of flight refuelling with the RAF.

With the closure of West Raynham, the Canberras of 100 Squadron arrived on 5 January 1976 at Marham, and the Canberras of 231 OCU arrived from Cottesmore on 18 February. With the disbandment of 98 Squadron at Cottesmore on 27 February, its Canberra E.15s joined 100 Squadron. The Canberras remained at Marham for six years, until they were all moved to Wyton, 100 Squadron moving in January 1982, followed by the much reduced 231 OCU in July.

Meanwhile the surprise invasion of the Falklands by Argentinian forces necessitated the full capability of AAR Victors, the first nine Victor K.2s being deployed to Ascension Island. They commenced operations on 20 April 1985, allowing Harriers and Sea Harriers to be flown direct from Britain to Ascension to be deployed aboard ship to the combat area. Probably the most demanding task was to support the Black Buck Vulcan bombing raids on the runway and military installations on the Falklands. The Victor K.2s returned to Marham in June 1985, continuing to support RAF combat operations, including in the Gulf War. The heavy use of the Victor K.2s resulted in their fatigue life being used up, and their eventual replacement by Brize Norton-based VC-10 tankers. The airframes with the highest remaining fatigue life were enough for a single squadron, with 232 OCU disbanding in April 1986, followed by 57 Squadron on 30 June. No. 55 Squadron continued to operate the Victor K.2s until they too were withdrawn and the squadron was disbanded on 15 October 1993. A number were allocated for preservation, including one at Marham, while the remainder were scrapped on site.

Improvements had started at Marham in 1977 including HAS sites with 12 shelters each on the southwest and eastern boundaries of the airfield to support two Tornado squadrons. Each HAS could accommodate a Tornado and included storage for up to four WE177 nuclear weapons. The first Tornado GR.1s arrived for 617 Squadron which re-formed on 1 January 1983, followed by 27 Squadron reforming on 1 May. Upon the Iraqi invasion of Kuwait on 2 August 1990, the Marham-based Tornadoes were deployed with the 55 Squadron Victor

Right: As part of the weapons fit for the Tornado GR.1s, the WE177 tactical nuclear weapon was withdrawn from use at midnight on 1 April 1994. *AWE Crown Owned Copyright*

Left: The Canberra PR.9s of 39 (1 PRU) Squadron arrived at Marham in December 1993 and were finally withdrawn when the unit disbanded at Marham on 28 July 2006. Canberra PR.9 XH131 at Marham on the day of disbandment with the 'C' Type hangars in the background. *Author*

K.2s to the Middle East until a return in March 1991, although deployments continued for patrols over southern Iraq while tension continued.

There was a requirement for the RAF to replace its Canberras, Phantoms and Jaguars with a more modern and capable aircraft, the unique Tornado GR.1A being developed to equip II(AC) and 13 Squadrons of the Reconnaissance Force at Marham. An initial total of 14 GR.1As were converted by BAe at Warton with the removal of two 27mm cannons, which were replaced by an electro-optical package based in IR technology resulting in the world's first filmless reconnaissance aircraft, the systems being operated by the occupant of the rear cockpit. In addition to the conversions, a further 16 GR.1As were built on the production line with the reconnaissance systems integrated by the RAF and the single development aircraft made its first flight on 11 July 1985. As production built up the first GR.1As were delivered to Bruggen and Laarbruch in April 1987, the aircraft having an all-weather day and night capability, although in heavy moisture there was some loss of quality.

With the run-down of RAF Germany, Tornado GR.1A-equipped II (AC) Squadron arrived at Marham from Laarbruch on 1 December 1991. With the withdrawal of the Buccaneers from service, both 27 and 617 Squadrons took over the maritime role, 27 Squadron being renumbered 12 Squadron in October 1993 and moving to Lossiemouth in late December. In April 1994, 617 Squadron also moved north to Lossiemouth. In February XIII Squadron began to fill the gap left by the departing squadrons when it arrived from Honington on 1 February 1994 with reconnaissance-dedicated Tornado GR.1As. At midnight on 1 April 1998 the WE177 tactical-nuclear weapon was withdrawn from service, and around this time XIII Squadron re-equipped with the much improved Tornado GR.4As, which carried SLIR and IR Linescan as part of their reconnaissance systems.

There remained a need for an improvement of the Tornado aircraft to counter the more capable Warsaw Pact defences, giving the Tornadoes an improved stand-off capability. The new version was to become the GR.4, and the original specification was very ambitious. With the ending of the Cold War, the requirement was much reduced and a development contract for a mid-life update (MLU) was issued to Panavia on 16 March 1989, with three aircraft converted as prototypes. The RAF had hoped for 26 new build GR.4s and conversion of 126 GR.1/1As at a cost of £1 billion, but with the collapse of the Soviet Union, the new-build batch was cancelled, and the conversion numbers reduced. Amongst the items cut out was the abandonment of the major structural modifications, and many of the systems improvements. The first development GR.4 was flown on 29 May 1993, but the programme was the victim of defence cuts, and it was shelved until after the Gulf War, with BAe receiving a contract on 14 July 1994 worth £640 million covering the conversion of just 80 aircraft, with options to convert 62 more from the year 2000. With the Options for Change and Front Line First defence Review, Tornado Squadrons were reduced to eight equipped with a wide range of different systems standards for the different roles. It was decided that the MLU would provide full commonality across the force apart from the specialist reconnaissance GR.4As. The very much downgraded MLU included the installation of a moving map display, GPS, enhanced weapon control system, a new wide-angle HUD with the standard symbology overlaid on the new FLIR image, with the cockpit made fully compatible with NVGs for improved low-level night attack. The first Tornado GR.4 to be delivered to an operational unit was to IX Squadron in Germany on 11 May 1998.

Meanwhile the Canberras had returned in the form of five PR.9s and two T.4s with 39(1 PRU) Squadron in December 1993. Their high-altitude performance made them invaluable for wet film reconnaissance over the Balkans and Middle East combat zones, until only three PR.9s were left in operation, when the squadron was disbanded at Marham on 28 July 2006.

With the new millennium approaching, preparations were made for the arrival of two Tornado bomber squadrons, the first being IX Squadron, which had been responsible for the original introduction of the Tornado to operational service, and was the first unit to be rearmed with the GR.4, arriving at Marham on 17 July 2001 from Bruggen. No 31 Squadron arrived in August, bringing Marham to the full strength of four Tornado squadrons, changing title from the Marham Reconnaissance Wing, to the Marham Wing. On 11 October 2004, Marham began preparations to become the BAE Systems-operated Depot Support Hub for all Tornado GR.4 support for the squadrons at Marham and Lossiemouth. At least two of the 'C' Type hangars were converted to a production line ready to start operations on 1 April 2005. In this complex, Tornadoes are stripped down to component parts for maintenance and modification to keep them in the front line of the RAF; the four Tornado squadrons form part of the 138 Expeditionary Wing, which provides full logistics support for the four squadrons when they are deployed overseas. With the threatened force reductions by the Government in October 2010, it is still not known what changes may be in store for the very active Marham RAF station, but the chances are that it will continue to operate four Tornado squadrons with the update and conversion line intact. It was later revealed that XIII Squadron would cease operations on 1 June 2011, leaving IX and 31 Squadrons in the strike role and 2 Squadron with the tactical reconnaissance duties.

Scampton, Lincs;
SK965795, 6 miles N of Lincoln, by A15

On 1 May 1958 the Dam Busters returned to Scampton with the reforming of 617 Squadron equipped with Vulcan B.1s, accommodated in the same hangar as the original Lancaster-equipped unit. The Vulcan B.1s were armed with WE177 free-fall nuclear weapons.

The Vulcan B.Mk.2 was very similar to the earlier version externally, but under the skin was a very different aircraft. Plans for an upgraded bomber were first studied when OR.330 was issued by the Air Ministry in 1954. This called for a very advanced aircraft capable of Mach 2.5 at an altitude of 60,000 feet with a still air range of 3,000 miles. This was not too far removed from what was to become the Concorde SST, and to cover the new bomber Specification R.156T was issued, with both Avro and Handley Page taking up the challenge. However, the infamous Defence White Paper of 1957 resulted in this requirement being shelved. Not to be discouraged, Avro proposed a range of improvements to the Vulcan, which was successfully sold as a concept to the Air Ministry and accepted by the Government. The main external difference was an increase in wingspan from 99 to 111 feet, while the leading edge shape was further revised to increase the buffet threshold to Mach 0.87. Instead of lateral and longitudinal control using ailerons and elevators, combined power-operated elevons were in four units along the wing trailing edge. To take advantage of the improved aerodynamics, more powerful Olympus 200 engines were fitted and new electrical systems were installed. Another visual change was the installation of ECM systems in an enlarged tailcone with the brake parachute located under the top of the fairing. The addition of these systems gave the AEO a much greater capability and the navigators had significantly improved equipment which reduced their workload.

The first Vulcan B.2 XH533 followed the earlier B.1s off the production line at Woodford and made its maiden flight on 19 August 1958 without the ECM tailcone, which was first fitted to XH534 in 1959. With the completion of development flying, the Vulcan B.2 was cleared for service in 1960 when 83 Squadron became the first unit to receive the new type at Scampton on 10 October 1960, followed by 27 Squadron on 1 April 1961 and 617 Squadron on 1 September the same year, concentrating the nuclear-armed Vulcan deterrent force at Scampton. The Vulcan force participated in many exercises with the USAF SAC, including weapons handling with the RAF Blue Danube, one of which could be carried, whereas two of the American nuclear weapons could be accommodated, as well as conventional iron bombs. Long range flights were conducted globally to practise navigation skills and operations deployed away from base.

With the increasing strength of Soviet air defences, it became obvious that sending bombers over enemy territory at any altitude with free fall weapons was suicidal, resulting in the need for a stand-off weapon. Work to develop such a weapon was shared between Avro and the Royal Aircraft Establishment

Left: No 83 Squadron was re-established at Scampton with Vulcan B.2s on 10 October 1960, with nine of their white-painted aircraft lined up on a disused runway. *RAF Museum P.021316*

Above: A Blue Steel-equipped 617 Squadron Vulcan B.2 on night standby at RAF Scampton. *IWM RAF T.4135*

(RAE) at Farnborough, and a development contract was issued by the Ministry of Supply in 1956. This rocket powered missile became Blue Steel requiring rocket fuel facilities and special handling equipment included dedicated transport vehicles. An additional complication was that Blue Steel was to be carried by both the Vulcan and Victor, which had very different ground clearance, and as it was too large to be carried internally, it was suspended half enclosed by modified bomb bay fairings, and the lower half exposed below the aircraft. The transport vehicle had to have a common interface with both aircraft for the airborne disconnection of hydraulic, cooling, heating and electrical supplies. Following a successful development programme, Blue Steel entered service with 617 Squadron at Scampton in September 1962, with 27 and 83 Squadrons receiving the modified aircraft during 1963 to become part of NATO's nuclear deterrent. The aircraft were on Quick Reaction Alert standby at Scampton and a number of deployed bases around Britain, where crews were ready for immediate action to comply with the notional four-minute warning, with Vulcans and Victors to be launched immediately Soviet ICBMs were detected coming over the horizon. The RAF maintained the primary nuclear strike deterrent until the Royal Navy took over the role on 30 June 1969.

Right: The Blue Steel air-launched stand-off nuclear missile required special ground handling vehicles to transport and load the missile into the Vulcan B.2s. The lower rear fin folded sideways to improve ground clearance, particularly with the lower fuselage of the Victors. *IWM RAF T.3612*

No 83 Squadron was disbanded on 31 August 1969, to be replaced by the returning 230 OCU in December to provide training for all Vulcan aircrew. Secure nuclear stores had to be built for the WE177 weapons, followed by special storage and maintenance buildings for the Blue Steel and its special fuel. A four-track ORP was constructed by the threshold at the north-east end of the main runway, with new dispersals on the north-west side of the airfield. With the gradual withdrawal of the Vulcan force, 27 Squadron disbanded on 29 March 1972, but re-formed on 1 November 1973 with Vulcan SR.2s in the MRR strategic reconnaissance role. When the Near East Bomber Wing disbanded at Akrotiri in Cyprus, 35 Squadron returned to the UK and joined the Scampton Vulcan Wing on 16 January 1975. 617 Squadron continued until 31 December 1981 when they were also disbanded, followed by 35 Squadron on 1 March 1982 and 27 Squadron on 31 March, when Vulcan operations ceased at Scampton. During the period of occupation by the Vulcans, the Hastings of the Strike Command Bombing School were also based at Scampton from 1 September 1972 until 1 January 1974.

With the departure of the V-Bombers, Scampton became the home of the Central Flying School (CFS) which included the Red Arrows RAF aerobatic team. In the mid 1990s under the Government's Front Line First programme, the CFS moved to Cranwell and Scampton was threatened with closure, but after much lobbying was reduced to C&M. However with the congested airspace at Cranwell, restricting the Red Arrows practices, the aerobatic team used the designated airspace at Scampton for training, in due course returning there and based in one of the hangars. There are plans for the Red Arrows to depart from Scampton in 2011 for Waddington, but again there could be restrictions on training at such a busy airfield, and the air above Scampton could still be busy with red Hawks. Flying does continue from the airfield with Hunters of Hawker Hunter Aviation, based on the north side of the airfield and providing facilities for contractor military training.

Waddington, Lincs;
SK985645, 4 miles S of Lincoln, between A15 and A607.

With the departure of the venerable Lincoln bombers and closure of Waddington in August 1953, the station was put under C&M for major preparations as a Class 1 bomber airfield. In addition to the lengthened main runway with its northern threshold close to the A.15 with an ORP for four V-Bomber rapid departures, dispersals and taxiways were strengthened and a number of specialist buildings were constructed with the nuclear stores across the eastern side of the A15. On 1 June 1955, the station reopened as a master diversion airfield and the first residents were the Canberras of 27 Squadron, which arrived from Scampton on 26 May 1955, followed by 21 Squadron on 1 June. Both units were only resident for around two years with 21 Squadron disbanding on 30 June 1957, and 27 Squadron disbanding on 31 December the same year.

Meanwhile the first Vulcans for the RAF were delivered to Waddington with 230 OCU responsible for Service Trials. The Vulcan was produced to the same demanding specification as the Victor, Air Ministry Requirement OR.229 defining the aircraft. The initial requirement was for a bomber capable of carrying the 'Blue Danube' nuclear weapon at a speed of 500 kts, and dropping the bomb from a height of 45,000 feet at a still air range of 3,500 miles. These figures were later increased to an altitude of 50,000 feet and a range of 5,000 miles. With this requirement passed to the Ministry of Supply, which issued Specification B.35/46 on 9 January 1947 to those companies believed capable of building such an aircraft. The bomb bay was to be capable of carrying any other weapons in development, as well as a reconnaissance pack. The primary navigation and attack system selected was the well proven, but with improved H2S, and the aircraft was to have a crew of five, consisting of two pilots, two navigator/bombardiers and a radio/ECM operator, all housed in a single pressure cabin.

With design work commencing under the project Avro Type 698, and ITP issued on 1 January 1948, it was decided to adopt a delta-wing planform, allowing plentiful space within the structure for systems, fuel and weapons carriage. With the first prototype nearly completed, the Government placed a contract on 14 August 1952 for an initial batch of 25 Vulcan B.1s, and VX770

Left: No 51 Squadron with three Nimrod R.1s moved from Wyton to Waddington in 1993, with the Nimrods due to retire in 2011. Nimrod R.1 XW665 of 51 Squadron in overall "hemp" colours on one of the few occasions these highly classified aircraft have been caught by the camera. *DTEO*

Left: With the retirement of the venerable AEW Shackletons, the Sentry AEW.1s entered service at Waddington in November 1990 with 8/23 Squadron. Sentry AEW.1 ZH107 of 8/23 Squadron was at the north end of the main runway at Waddington in August 1996. *Author*

made its maiden flight from Woodford on 30 August. Power of the Vulcan was to come from four Olympus turbojet engines, but to begin testing the aircraft as soon as possible the first prototype was initially powered by Rolls-Royce Avons, the second prototype making its maiden flight with Olympus engines on 3 September 1953. The Vulcan was to be highly manoeuvrable, but early testing showed that the simple delta-shape wing with a straight leading edge caused compressibility problems, the correction being to incorporate a new wing leading edge with compound sweep. The success of the Vulcan development programme was such that the Ministry increased the orders to 45 of the Vulcan B.1s, later to be followed by the improved B.2.

With the development almost complete, the first Vulcan delivery to the RAF was XA895 on 20 July 1956, although it was soon replaced by XA897, with 230 OCU at Waddington. On 2 September XA897 departed Britain for a promotional tour of Australia and New Zealand, but on its return, crashed at Heathrow in bad weather when it undershot the runway, killing the four occupants in the rear of the cabin; the two pilots ejected successfully, including AM Sir Harry Broadhurst, C-in-C Bomber Command. One of the major points of concern was the inability of the rear crew members to be able to escape from the aircraft in an emergency, as only the two pilots had ejector seats. When 230 OCU had completed the service trials and sufficient crews had been trained, the first regular RAF unit to equip with the Vulcan was 83 Squadron at Waddington on 20 May 1957 and once the crews had worked up on the type, two Vulcans were entered for the US Strategic Air Command Navigation and Bombing competition at Pinecastle AFB. This competition later developed into "Giant Voice" and eventually the RAF Vulcan crews were so successful at winning most of the trophies, that the USAF changed the rules.

Meanwhile back in Britain, 101 Squadron re-equipped with Vulcan B.1s at Finningley in October 1957 and 617 Squadron received theirs at Scampton in May 1958. With the development of the much improved Vulcan B.2, 29 of the remaining 34 B.1s were modified to as close as possible to the B.2 standard, which included ECM systems in an enlarged tailcone and the fitting of in-flight refuelling probes in the nose. The new version became the B.1A and to simplify the logistics, the B.1As and B.2s were grouped together with Waddington becoming the home of 44, 50 and 101 Squadrons with Vulcan B.1As in the conventional role, with the B.2s taking the nuclear

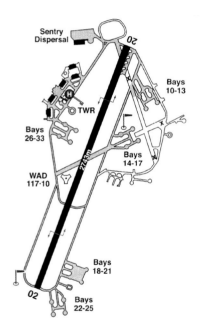

Above: Waddington reopened as a reserve station in October 1926 and was developed in the mid 1930s within the RAF Expansion Scheme with five 'C' Type hangars located within the technical site on the western boundary of the airfield. Bomber operation started in May 1937, but the airfield was closed from May until November 1943 for the construction of three runways. The station was then the base for RAAF Lancaster bomber squadrons until the final operation in April 1945. After World War 2 the station continued with RAF bomber squadrons operating Lincoln B.2s until it was closed in August 1953 for preparations to accept the V-Bomber force. Amongst other improvements, the main runway was extended to 9,000 feet and the other two runways disused. The first Vulcans arrived with 230 OCU in July 1956, and remained until 50 Squadron disbanded in March 1984. Waddington is now used by AWACS-based in a new facility at the north end of the airfield, together with other intelligence gather units. *Pooley's Flight Guides*

Left: No 50 Squadron re-equipped with Vulcan B.2s at Waddington in January 1966, and finally operated converted Vulcan K.2s from September 1982 until disbanding on 31 March 1984 as the final RAF Vulcan unit. Vulcan B.2 XL427 in the matt wraparound camouflage departs Waddington on 30 May 1981. *Author*

deterrent role. The B.1As were withdrawn from service by 1967 to be replaced by the B.2s. Meanwhile 44 Squadron re-formed at Waddington with Vulcan B.1s on 10 August 1960, to be joined by 101 Squadron from Finningley on 26 June 1961, and 50 Squadron completed the Waddington Wing with B.1s when it re-formed on 1 August 1961. The conversion to Vulcan B.2s commenced with 50 Squadron in January 1966, followed by 44 Squadron in September, and 101 Squadron in December 1967. The Vulcans were withdrawn from the nuclear-deterrent role on 30 June 1969 when the RN Polaris submarines took over. When Turkey invaded Cyprus, 9 and 35 Squadrons returned to the UK, 9 Squadron arriving at Waddington with B.2s on 15 January 1975.

In the early 1980s the retirement of the Vulcan force commenced and at Waddington 9 Squadron was in the process of disbanding when Argentine forces invaded the Falkland Islands on 1 April 1982. To achieve the required endurance the Vulcans would need the long-redundant flight-refuelling system reinstated, allowing the hard-pressed Victor tanker fleet to replenish the Vulcan tanks during the long flight south from Ascension Island to the Falklands. Ten of the late production were chosen for operational use, with six selected for the strike force; the aircraft had to be adapted for the dropping of conventional iron bombs and a suitable inertial navigation system had to be fitted. The selected crews had to relearn weapons delivery and flight refuelling skills. With the preparations complete, the first two Vulcans, XM598 and XM607, departed Waddington on 29 April for Ascension Island off the west coast of Africa. The first of what was known as the "Black Buck" raids was launched on the night of 30 April/1 May and after eleven in-flight refuellings XM607 dropped 21 bombs diagonally across the runway at Stanley. Following a series of Black Buck raids, the Vulcans returned to Waddington on 14 June, the day of the Argentine surrender. No. 9 Squadron had formally disbanded on 1 May 1982 and was followed by 101 Squadron on 4 August and 44 Squadron on 21 December 1982, ending the career of the Vulcan as a bomber. However, the high utilisation of the Victor tankers in the Falklands campaign had used up valuable fatigue life, and in the short term, pending the entry into service of the VC-10 tanker conversions, six Vulcans were converted at Woodford to flight refuelling tankers with a single hose-and-drogue unit (HDU) under the tailcone, and an additional 8,000lb bomb-bay fuel tank fitted. All six conversions were completed between August and September 1982 and were issued to 50 Squadron at Waddington in addition to four standard bombers to be used for training. With the VC-10s requiring the HDUs, 50 Squadron was disbanded on 31 March 1984 and the last Vulcan was withdrawn from service; after a period as a display aircraft, Vulcan XH558 was retired to Bruntingthorpe and was later to be restored to flying condition.

Waddington then had a change of role, being earmarked for the Nimrod AEW.3, but with its cancellation the joint 8/23 Squadron re-formed with the E-3 Sentry AWACS aircraft in November 1990. In 1993 51 Squadron moved to Waddington from Wyton, with Nimrod R.1 intelligence-gathering aircraft, which are due to be withdrawn in 2011. In due course they are to be replaced by ex USAF EC-135s from desert storage and of greater age than the Nimrods. On 1 April 2004 5 (Army Co-operation) Squadron re-formed at Waddington in the battlefield and ground surveillance role in support of the Army and received the first of the five Sentinel R.1s on 1 December 2008. The Sentinels are active in Afghanistan in support of the ground troops where their sensors can, amongst other intelligence targets, detect potential IEDs (improvised explosive devices). In its defence review of October 2010, the Government has decreed that 5 Squadron will disband after the removal of British troops from Afghanistan.

Wittering, Cambs;
TF045025, 3 miles SE of Stamford, by A1

The preparations for the V-Force included the lengthening of the runway to 9,052 feet towards Collyweston, with 'H' layout four aircraft dispersals at the western end to give rapid access to the runway, and a four track ORP at the eastern end of the runway. The Bomber Command Development Unit (BCDU) arrived in July 1954 and 1321 Fight formed to carry out dropping trials from a Valiant of inert rounds of 'Blue Danube', until the Flight disbanded on 15 March 1956. At the beginning of July 1955, Wittering became a prime V-Bomber base with the arrival of the Valiants of 138 Squadron, which was joined by 49 Squadron on 1 May 1956. In the autumn of that year 138 Squadron deployed to Malta, carrying out conventional attacks on Egyptian airfields during the Suez Crisis. Meanwhile 49 Squadron was allocated to undertake nuclear weapons trials at Christmas Island during 1956-57. The BCDU wound down in March 1960, and 7 Squadron brought its flight-refuelling Valiants from Honington on 1 September, bring the Wittering Wing up to the full strength of three squadrons. With the discovery of structural problems with the Valiants, the type was withdrawn from service and 49 Squadron disbanded on 25 June 1961, 138 Squadron on 1 April 1962 and finally 7 Squadron on 30 September, ending the Valiant service at Wittering.

The Victor B.1A had been a modest improvement on the original B.1, the most obvious change being the installation of ECM systems, similar to the Vulcan, within an enlarged tail cone. Despite this there was still a need for an improved performance aircraft, and as development of the Sapphire engines was stopped, the increased power would have to be from a new engine. As a result the 17,250lb-thrust Rolls-Royce Conway was selected and the wingspan increased by 10 feet giving a service ceiling of 60,000 ft. The initial order for 18 aircraft was placed in January 1956, the aircraft being designated Victor B.2 with many other improvements to the operational and defensive systems. There was no prototype as such, the first production aircraft being XH668 which first flew on 20 February 1959 (but was lost on 20 August in Cardigan Bay), the development trials being handled by early production aircraft. From September 1960 two Victor B.2s were allocated to Avro at Woodford for Blue Steel trials, later moving to Boscombe Down and then Woomera in Australia for live firing trials. The final production of the Conway developed 20,600lb thrust.

Wittering was chosen to become the base for the Victor Blue Steel Force and it was not long before V-Bombers returned to Wittering, with the Victor B.2 Trials Unit forming the nucleus of 139 Squadron which had re-formed at

Above: Britain's first atomic bomb, the 'Blue Danube', was delivered to Wittering on 7 November 1953 in great secrecy. There were no aircraft capable of carrying the weapon until the Valiants of the BCDU arrived in July 1954, and carried out trials until March 1956. *AWE Crown Owned Copyright*

Wittering on 1 February 1962. On 1 May 100 Squadron was re-formed at Wittering, also with Victor B.2s, making the station the sole operational base for the nuclear-deterrent Blue Steel stand-off cruise missile-carrying Victors. The facilities at the station were developed to handle not only the new Victors, but the special requirements of the rocket-powered Blue Steel. The Victor B.2 Training Flight formed in March 1964 followed by the Navigational Bomb Site Development Unit (NBSDU) and the Bombing &

Right: No 138 Squadron were responsible for the introduction of the Valiant B.1 into RAF service and moved from Gaydon to Wittering on 6 July 1955. The squadron operated Valiant B.1s from February 1955 until March 1962, the overall reflective white painted XD861 being on the squadron strength. *RAF Museum P004470*

Right: Harrier T.4 V:20 Squadron/233 OCU inside the 'Gaydon' hangar at Wittering on 5 October 1994. *Author*

Navigation Systems Development Squadron (BNSDS) on 1 August 1966. With the decision to operate Victors at low level, the aircraft were painted in camouflage on the top surfaces, retaining the reflective white on the undersides, but care had to be taken not to overstress the aircraft due to strong turbulence nearer the ground. When armed with the Blue Steel missile, the Victor B.2s approached the hostile area at low level and an initial climb 25 to 30 miles from the target, and released the missile, which would then climb to 71,000 ft before diving onto the target. With both squadrons declared operational, a pair of fully-armed aircraft would be on immediate alert round the clock, allowing the earlier Victor Mk.1s to be released for conversion to tankers. Because of its low-level strike range, Blue Steel was phased out with 100 Squadron disbanding on 30 September 1968, followed by 139 Squadron on 31 December, the Victors being modified at Woodford for the flight-refuelling tanker role, and going to Marham.

On 1 February 1969 Wittering was transferred to Air Support Command with the first aircraft being Whirlwind HC.10s of 230 Squadron as part of 38 Group, the squadron remaining until moving to Odiham on 1 January 1972, by which time it had re-equipped with Pumas. Another element of 38 Group were the Hunter FGA.9s of 45 Squadron which arrived from West Raynham on 29 September 1972, with 58 Squadron reforming as an addition on 1 August 1973. On 2 September 1974 the two squadrons formed a Hunter Wing for both training and an operational ground-attack role until disbandment on 26 July 1976.

Below: With the arrival of 138 Squadron and the Valiants on 6 July 1955, Wittering became the first major strategic V-Bomber RAF station. The squadron maintained a round-the-clock readiness and operated Valiant B(PR).1s from March 1956 until May 1961, the squadron disbanding on 1 April 1962. *RAF Museum P004475*

Work on the unique Harrier, then known as the Hawker P.1127, started at Kingston in mid-1958; it was powered by a revolutionary vectored-thrust jet engine, later to become the Bristol Siddeley Pegasus. In March 1959, the Hawker Board decided to go ahead with building a prototype as a private venture, but a month later the RAF issued GOR.345, written around the P.1127 as a Hunter replacement to enter service in 1965. In June 1960, the Government provided funding for two prototypes, and the first, XP831, was delivered from Kingston to the Hawker flight development airfield at Dunsfold on 15 July 1960, with the first of six Pegasus engines arriving in September. As the early Pegasus engines lacked in power, the lightly loaded prototype made its first tethered hover flight on 21 October, with the first conventional take-off on 13 March 1961. At the end of 1960, the Government ordered four more prototypes in a development batch, and the second aircraft, XP836, took to the air on 7 July 1961. On 12 September 1961 Bill Bedford, the Hawker Chief Test Pilot, made the first transition from vertical to horizontal flight, and the first deck landing was made on HMS *Ark Royal* in February 1963. With flight development continuing on the first six aircraft, although some were lost in accidents, the improved Kestrel was proposed and nine were ordered for a tri-national evaluation squadron, sponsored by Britain, America and West Germany, the aircraft being powered by a 15,500lb-thrust Pegasus engine. The first of nine Kestrels flew on 7 March 1964 and the Tripartite Evaluation Squadron (TES) began its evaluation at West Raynham on 1 April 1964, where operations were made from dispersed sites. A total of 930 sorties were flown over 600 hours with the loss of one Kestrel during the eight months of the programme, the operation being considered a total success. Although the Germans showed no further interest, the US Marine Corps were to become great supporters of the concept.

A new RAF requirement, ASR.384, was issued in February 1965 to define what was known as a revised P.1127 (RAF) powered by a combat-capable Pegasus 101, an order for six aircraft being placed. This resulted in a 90 percent redesign of the airframe and a zero-zero Martin Baker ejection seat was installed to allow the pilot to abandon in most regimes of operation. The new aircraft had many new systems incorporated for its ground-attack role against enemy armour, supply convoys and troop concentrations. Operating from deployed sites rather than easily targeted runways would make the location of the Harriers more secure, as well as being placed closer to the front line. Amongst the aids was a head-up-display (HUD), to allow the pilot to keep on to the target, while also seeing his flight parameters, without looking

Above: A Wittering Wing Victor B.2 carrying an inert Blue Steel missile flies over Abingdon on 14 June 1968. *Author*

down at the instrument panel. The first of the six development Harriers, as the aircraft had been named, was flown on 28 December 1967, following the placing of the initial contract for 60 Harrier GR.1s at the beginning of the year. These 60 aircraft were to equip one UK and one Germany-based squadron, but subsequent orders were placed for 118 first-generation Harrier GR.1s for the RAF.

Meanwhile the aircraft and operations for which Wittering has been famous for in over 40 years is the V/STOL Harrier. The Harrier Conversion Team (HCT) began its training at Dunsfold on 1 January 1969, and the first Harrier was delivered to Wittering on 18 April. The HCT became the Harrier Conversion Unit (HCU) on 1 April 1970. On 18 July 1969, 1 Squadron re-formed, initially with some Hunters, the task being to explore the Harriers potential for operational service, and to take advantage of its unique capabilities. As part of this assessment, 1 Squadron took nine aircraft to Akrotiri in Cyprus for the first detachment in March 1970. This was followed by "off base" exercises around Wittering airfield and a number of local locations, and in July 1(F) Squadron had received its total compliment of 18 Harrier GR.1s, declared fully operational in September. It was deployed to Norway for the defence of NATO's northern flank, with many more exercises to follow to take the best advantage of this unique aircraft.

The initial pilot conversions for 1 Squadron, and later 3 and 4 Squadron had been undertaken with single seat Harrier GR.1s, with an instructor flying in a Hunter. There was a need for a two-seat version to better demonstrate the special flight characteristics of the Harrier, and a year after the initial Harrier order, the RAF ordered two development two-seaters. The

On 1 October 1970 the HCU became 233 OCU and adopted the markings of 20 (R) Squadron, as shown on Harrier GR.1 XV797 in a 'C' Type hangar at Wittering on 22 October 1970. *Author*

forward fuselage was lengthened to allow the installation of a raised tandem cockpit for the instructor behind the pilot under training. The first Harrier T.2 made its maiden flight from Dunsfold on 24 April 1969, but was lost six weeks later with a fuel system problem, but fortunately the second aircraft flew on 14 July. The first of 12 production T.2s flew on 3 October 1969, and deliveries began to the HCU at Wittering in July 1970. The HCU became 233 OCU on 1 October 1970 with the role of training pilots in the operation of the Harrier prior to joining a front-line squadron. With the overall reduction of RAF squadrons in operation, some units took squadron titles as reserve squadrons, and 233 OCU became 20 (Res) Squadron on 1 September 1992, to be renumbered 4 (Res) Squadron on 31 March 2010 due to defence cuts. In the event of conflict, these squadrons would become active, staffed by the instructors.

Harrier development was always driven by the ever increasing power made available from continual improvements to the Pegasus engine. The first relatively modest update was the retrofitting of 41 Harrier GR.1s with the 20,500lb thrust Pegasus 10 Mk.102 with the aircraft being re-designated Harrier GR.3s. A further increase in thrust of 1,000lb was achieved by the Pegasus 11 Mk.103, and was fitted to 40 new production Harrier GR.3s, with 12 more ordered in 1974.

Both 1 Squadron and 233 OCU initially operated Harrier GR.1s together with the two-seat T.2, with the single-seat version replaced by the GR.3. In October 1975 six Harrier GR.1As of 1 Squadron were detached to Belize to defend against invasion threats from Guatemala. What was expected to be a short deployment ended up lasting 18 years, as the specially created 1417 Flight with four Harrier GR.3s which was finally withdrawn in 1993. When Argentine forces invaded the

Left: Harrier T.10 ZK658 106:1 Squadron lifts off from Fairford on 18 July 1997. The two-seat T.10s were used by 1 Squadron for continuation training at Wittering. *Author*

Falkland Islands on 1 April 1982, 1 Squadron Harrier GR.3s and FAA Sea Harriers embarked on HMS *Ark Royal* for the defeat of the aggressors and recapture of the Falklands. During the conflict, 1 Squadron Harriers flew some 130 sorties for the loss of three aircraft, the pilots ejecting successfully.

The greatest improvement of the Harrier concept was the US Marines-sponsored AV-8B Harrier II following the success of the original AV-8A where licence manufacture was planned to be undertaken for the US Marines by McDonnell-Douglas (McAir). Although US manufacture of the AV-8A was not to happen, when it came to the major development which was virtually a new design based on the original concept, it became a US Government lead programme. In August 1981 a Memorandum of Understanding (MoU) was signed between the British and American Governments confirming an initial order for 60 Harrier GR.5s for the RAF. Because of much larger orders from the US Marines of up to 336 aircraft, McAir would have 60 percent of man-hours on airframe manufacture with BAe taking the balance, while Rolls-Royce were allocated 75 percent of Pegasus work, with Pratt & Whitney responsible for 25 percent. With a more powerful and larger diameter Pegasus 105 developing 21,750lb thrust, the Harrier GR.5 fuselage was redesigned with new systems, and there was a new wing to provide an additional 50 percent fuel capacity and greater weapons load carrying capability on seven external pylons.

Flight development of the Harrier II began with hovering tests at St Louis on 5 November 1981 by the first of four development aircraft, followed by 12 pre-production Harrier IIs to get the production line established with the first aircraft delivered to the US Marine Corps on 12 January 1984. Externally similar to the AV-8B, the Harrier GR.5 introduced many changes internally to meet the RAF needs, particularly the integration of British systems, equipment and weapons, including self-defence AIM-9G Sidewinders. Although the first Harrier GR.5 was delivered to Wittering on 1 July 1987, it did not make its first operational flight until 30 March 1988; further deliveries were made from May, with 1 Squadron having eight on strength by the end of the year. The Harrier Conversion Team was re-established in February 1987 with the first 233 OCU course starting on 18 July and all 1 Squadron pilots converted by the end of April 1989, and the unit was fully operational by the autumn of 1989. The conversion of the RAF Germany squadrons continued with 233 OCU with the GR.3 being finally withdrawn on 7 December 1990. The much improved GR.5 was used by 1 Squadron on Operation Warden, which was policing Iraq's northern border areas and operating from Incirlik in Turkey from August 1993 to April 1995.

The next Harrier development was the GR.7 which was provided with a night attack capability, which was preceded at Kingston by 19 interim GR.5As. These were later converted to the full GR.7 standard including the updating of 58 GR.5/5A Harriers. The first of the new build GR.7s made its maiden flight on 29 November 1989, the major improvements being digital moving maps, forward-looking infra-red (FLIR) and a night-vision goggles (NVG) compatible cockpit. This allowed the Harriers to undertake longer-range interdiction against enemy armour re-enforcements behind the forward battle area under the cover of darkness. The pilot also had the benefit of HUD to keep himself current on the flight data without having to look down. With the GR.7 development continuing, the GR.5 was gaining popularity with the pilots. Having fired their first AIM-9L air-to-air missiles at a Valley armament practice camp (APC) in July 1989, the aircraft had their camouflage adapted with white splodges for a deployment to Bardufoss in Norway, which was repeated in March and April 1990. Additional orders for the Harrier GR.5s brought the overall total to 96, of which the last 27 were built as GR.7s.

With the initial deliveries of the GR.7s to the potential front line in Germany, it was not until 2 June 1992 that the first Harrier GR.7s arrived at Wittering for 1 Squadron. The first GR.7 sortie was flown on 2 June and the squadron was fully equipped by November of that year. After the rundown of 20 Squadron Tornado GR.1s at Laarbruch in May 1992, their number plate was passed to 233 OCU, becoming 20 (Res) Squadron. On 30 January 1995 20 Squadron received the first of the Harrier T.10s with similar handling characteristics to the GR.7, a total of seven delivered, and one to 1 Squadron. In August 1995, 1 Squadron, by now fully equipped with Harrier GR.7s, operated from Italy in Operation Deny Flight over Bosnia, and after carrier deck training, the squadron embarked on HMS Invincible in support of Operation Bolton, to ensure that Iraq was complying with the UN chemical weapons restrictions. In early 1999 16 Harriers of 1 Squadron together with part of 20 (Res) Squadron were back in Italy for Operation Allied Force, which was the air assault on Serbia as part of the Kosovo campaign. The Harriers flew an average of 20 strike sorties a day, dropping precision-guided Paveway weapons against power stations, bridges, ammunition storage and similar military and logistics targets.

What has become the ultimate development of the Harrier is the GR.9/9A and T.Mk.12, which include the two major features of the Integrated Weapons Programme (IWP) and the installation of the 23,800lb thrust Pegasus Mk.107 engine with a conversion line of GR.7s set up at Cottesmore. In April 2002 BAE Systems received an interim contract for the development of the GR.9 aircraft, with a further £150 million contract in May 2003, the first conversion taking to the air in May the same year. This programme of conversion and update was about complete when the Harriers were withdrawn, when the type had gained its greatest potential.

On 1 April 2000 the decision was made to form Joint Force Harrier consisting of 1 Squadron, 4 Squadron, 800 NAS and 801 NAS to be based at Cottesmore. No 1 Squadron therefore left Wittering on 28 July for the short flight north to its new base. Wittering then became solely responsible for the full training of RAF and RN Harrier pilots, engineers and maintainers in support of Joint Force Harrier. The Government strategic defence review of October 2010 announced the premature retirement of the Harrier force, which by then was operating the ultimate Harrier GR.9, with 1, 4 and 800 Squadrons disbanding at Wittering on 28 January 2011. RAF Wittering therefore closed to flying, but is unlikely to be disposed of by the MOD, as when the site was originally acquired by the government, a condition was that when no longer required by the military, it should be offered to the original owner in the state in which it was acquired. To do this would be very expensive, and is probably the reason that such an old RAF station has been in the front line for so long.

Wyton, Cambs;
TL285741, 4 miles NE Huntingdon, alongside the A141.

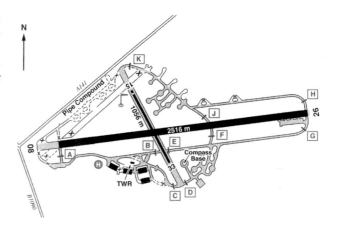

Before the arrival of the V-Bombers the strategic reconnaissance role was commenced with Canberra PR.3s of 540 Squadron, which arrived from Benson on 26 March 1953 to join the re-equipped 58 Squadron, which exchanged its Mosquitos for Canberra PR.3s in December of the same year. To demonstrate the long-range rapid deployment capability, in June 1953 three Canberra PR.3s flew newsreel footage of HM The Queen's Coronation across the Atlantic to Canada. In October Canberra PR.3 WE139 won the high-speed section of the London to New Zealand Air Race, the aircraft now being preserved in the RAF Museum. On 15 May 1954, 542 Squadron re-formed at Wyton with Canberra PR.7s, which were able to operate up to 50,000 feet, but the squadron disbanded on 1 October 1955. It re-formed on 1 November from 1323 Flight with Canberra B.2s and B.6s prior to moving to Weston Zoyland on 15 December 1955. Meanwhile 540 Squadron received Canberra PR.7s in June 1954 with 58 Squadron receiving theirs in January 1955.

On 18 November 1955 Valiant B(PR).1s arrived at Wyton with 543 Squadron, later having the flight refuelling capability added. On 31 March 1956 540 Squadron disbanded, with 58 Squadron remaining, converting to Canberra PR.9s in January 1960. Victor B.1s first arrived in April 1958 for the newly-formed Radar Reconnaissance Flight (RRF), the aircraft being equipped with "Yellow Aster" high-definition sideways-looking radar, and the RRF disbanded on 1 September 1961. Although Wyton was primarily a strategic reconnaissance base, the squadrons also undertook large-scale surveys and mapping tasks which particularly suited the Valiants, one of the tasks being a full-scale survey of Thailand. The Canberras of 58 Squadron undertook a 200,000-square-mile survey of East Africa in 1960. Early in 1961 543 Squadron Valiant made a survey from 40,000 feet of the island of Tristan da Cunha, followed later in the same year by recording the hurricane damage in British Honduras.

With the closure of Watton, the highly classified Comet 2Rs of 51 Squadron arrived at Wyton on 31 March 1963, together

Above: Wyton was selected as a bomber station within the RAF Expansion Scheme with four 'C' Type hangars located on a technical site to the south of the airfield. Construction work started at the end of 1935 and was mainly completed by November 1936. From August 1942, three runways were built with perimeter tracks and 36 heavy-bomber dispersals. The 3 Group Pathfinder Force had been formed at Wyton in August 1942 and continued until the end of World War 2. The station continued in the bomber role with Lincoln B.2s from November 1946 until September 1950. Wyton was selected for the strategic reconnaissance role in 1952 and the main runway was extended to 9,000 feet, with the first reconnaissance Canberras arriving from Benson in March 1953. Valiants arrived with 543 Squadron in November 1955 but were withdrawn in December 1964 due to structural metal fatigue. The replacement Victor SR.2s arrived in May 1965 and remained in service until April 1975, when they were withdrawn. Wyton then became the major RAF Canberra base until the airfield closed in November 1993 with the withdrawal of most of the Canberras. Flying restarted in September 1999 with the University Air Squadrons and Air Experience Flights. *Pooley's Flight Guides*

with specially equipped Canberra B.6s. Three of these were on the squadron strength and were the first of the Comet 2s built originally for BOAC. They were flown unpressurised and retained the original square windows, and with the various dielectric fairings in the fuselage recorded radio and radar frequencies of Warsaw Pact and Soviet defence systems, to allow countermeasures to be devised.

The structural problems with the Valiants were first discovered on a 543 Squadron aircraft, despite them operating in a much less demanding environment, and resulted in the withdrawal of the type from the squadron in December 1964. The Victor B.2 had 40 percent greater range than the Valiant, and a 15 percent increased ceiling, studies for which had started in October 1961, with the first aircraft withdrawn for conversion to the B(SR).2 role in 1964. The aircraft did not have the full EW jamming systems since it was not expected to operate over hostile territory, but was required instead to undertake high-level maritime reconnaissance sorties or survey work. The bomb bay was therefore reconfigured with

Left: RAF Wyton c.1960 from the air shows the four 'C' Type hangars, with the air traffic control in front of the centre hangar with a repaired roof following a fire. Behind the hangars is the main technical site with 'H' barrack blocks to the right and married quarters in the background. RAF Wyton was closely linked with nearby RAF Brampton where the reconnaissance results were processed and analysed. *RAF Museum P017589*

additional fuel tanks at the front and rear, and a camera pallet in between, which could be equipped with a range of cameras for day and night operations.

After a short wait the much more capable Victor B.2(SR)s were supplied to 543 Squadron from 18 May 1965 and amongst their tasks was a survey of Saddleworth Moor in an attempt to locate the Moors murder victims. By the late 1960s high-level reconnaissance had become vulnerable to missile attack, and therefore 58 Squadron changed to the low-level PR role before disbanding on 30 September 1970, its place being taken by 39 Squadron with Canberra PR.7s until they were replaced by PR.9s in 1972. The emphasis on intelligence gathering was ever increasing, resulting in the establishment of the Electronic Warfare and Avionics Unit at Wyton to manage the installation of the special classified systems in the aircraft. In July 1971 the first Nimrod R.1 arrived at Wyton for 51 Squadron to replace the venerable Comets, the last of which was retired in January 1975. Following fitting-out with the electronic systems, the first Nimrod R.1 made its first operational flight on 3 May 1974 and in 1976 51 Squadron gave up their last Canberra B.6, WT305. With the disbandment of 543 Squadron, their last Victor left Wyton on 3 April 1975.

Canberra operations continued at Wyton with the joint RAF/RN 360 Squadron from August 1975 with Canberra T.17s electronic warfare trainers to give defensive radar operators, both on land and at sea, a chance to practise their skills. A total of 24 Canberra B.2s were converted to the T.17 configuration, with the previous clear bomb-aiming nose replaced by a rather lumpy bulbous nose housing the electronic systems. On 3 October 1978 Canberra PR.7s with 13 Squadron arrived from Malta, their role being low-level graphy, while 39 Squadron Canberras had been equipped with Infra-red Linescan to allow night operations. With the disbandment of 13 Squadron on 1 January 1982 and to ease the logistic support, Wyton became the base for all RAF Canberras. This resulted in 100 Squadron moving in during January to take over 13 Squadron Canberra PR.7s, and also absorbed the target-towing 7 Squadron. No. 100 Squadron operated a wide variety of Canberras, providing target facilities for the Tornado F.3s with 229 OCU at Coningsby, the B.2s as "silent targets", while the PR.7s and E.15s were used for radar station calibration purposes.

Left: No 543 Squadron received Valiant B(PR).1s from 18 November 1955 and continued to operate these aircraft until they were withdrawn due to structural problems in December 1964. Valiant B(PR).1 WZ396 was one of the 543 Squadron aircraft, and was seen over the boundary fence on its dispersal at Wyton. *Author*

Above: On 9 July 1968, 51 Squadron, then operating Comet R.2s, was presented with its standard at RAF Wyton. Two of the squadron's three Comet R.2s flanked the ceremony, with the 'C' Type hangars and main technical site of the station beyond. *RAF Museum P019775*

The Canberra TT.18s could tow a banner target at the end of a 900ft line for safe air-to-air firing, two of the aircraft being detached to Cyprus for the regular armament camps. For ground-to-air target practice with Rapier missiles, two Rushden targets could be trailed up to five miles behind the aircraft. On 29 July 1982, the Canberra compliment was completed by the arrival from Marham of a flight of Canberra B.2s and T.4s of the much-reduced 231 OCU, bringing over 40 Canberras to the station. When 39 Squadron disbanded on 28 May 1982, it was replaced by No.1 graphic Reconnaissance Unit with Canberra PR.9s, one of their roles including medium-level survey mainly for mapping purposes, both military and civil. On 6 September 1991 the first of a dozen Hawks was delivered to 100 Squadron, commencing the replacement of the Canberras, which was completed by 1 January 1992. After some 41 years of training Canberra aircrews, 231 OCU disbanded on 23 April 1993, its task finally having been completed. The end of the Canberra service at Wyton came on 31 October 1994 when 360 Squadron disbanded, with all its Canberras, apart from one to be scrapped on site. The one survivor was B.17 WD955, which made the last flight on the day of disbandment, as the longest-serving frontline RAF aircraft, and was flown to preservation in Norway. The final Canberra departure was 39 (1 PRU) Squadron to Marham in November 1993, followed by the Nimrod R.1s of 51 Squadron which moved to Waddington. Wyton closed to flying.

In April 1994, Wyton merged with Brampton, and later Henlow to become RAF Brampton, Wyton, Henlow, but the airfield was reactivated in September 1999 as home for the Tutors of the Cambridge and London University Squadrons (UAS) and 5 AEF (Air Experience Flight). By the entrance to the airfield is a preserved Canberra PR.9 to remind people of the past glories of the station.

Chapter 8

Maritime Reconnaissance

With Britain being an island nation, not only has the surrounding water been a natural barrier to invasion, but because we are not self-sufficient in many of our vital supplies, we have relied on maritime reconnaissance to protect our shores since aviation became a practical mode of operation. This is not just for defence, but also for fishery protection, the control of illegal immigration, ocean rescue patrols and in more recent years the protection of our oil rigs in the North Sea. It only needs Russian nuclear submarines to park off our major ports, and we would be brought to our knees within about a week. The premature withdrawal of the Nimrods at Kinloss in March 2010, followed by the cancellation of the replacement Nimrod MR.4 in October, just when it was practically paid for after an enormous investment, leaves Britain without an effective means of protecting its coastline. There is no other major coastal nation in the world which has so neglected its maritime defence, especially after investing so much money in a very expensive defence programme. Cold War maritime reconnaissance tended to be located in both the south and north of Britain after World War 2, but as the Cold War progressed, it became more logical to protect the northern approaches, where Soviet nuclear submarines were departing from around the North Cape and sailing submerged down the Iceland gap, out into the North Atlantic. It was rumoured that there were so many recording devices on the sea bed that every move could be detected and identified.

Kinloss, Moray;
NH070630, 9 miles W of Elgin, by B9089

Kinloss was built in 1938/39 as a late RAF Expansion Scheme station for the training role; with the technical site located on the south side of the landing area which included three 'C' Type hangars and a mixture of permanent and temporary buildings. The first unit was 14 SFTS with Oxfords and Harvards, but in mid-April 1940 the station was transferred to Bomber Command and detachments of 77 and 102 Squadrons with Whitleys were resident for less than a month. Meanwhile on 15 April 1940 45 MU formed, with its main site consisting of three 'L' hangars located on the northern boundary of the airfield, with a total of 12 hangars dispersed around the site. On 17 May 1940, 19 OTU was formed for night bomber training which continued until the end of the war, with a secondary role as a reserve bomber unit should there be an invasion threat from the north. In 1942 Kinloss was chosen as an advanced base for Norwegian operations and brought up to full heavy-bomber standards with three metalled runways, the Tirpitz being finally sunk by Lancasters of 9 and 617 Squadrons flying from Kinloss in November 1944. The OTU closed on 26 June 1944, its work completed, and the station was taken over by Coastal Command, 6 OTU training Beaufighter and Mosquito crews. It was redesignated 236 OCU on 31 July 1944, and was responsible for training the crews of maritime Lancasters, Neptunes, Shackletons and finally Nimrods. No 45 MU closed during 1946 after scrapping many war-surplus aircraft.

From 14 December 1950, 120 Squadron flew Lancasters on maritime patrols and converted to Shackleton MR.1s in March 1951 with a move to Aldergrove in Ulster on 1 April 1952. While at Aldergrove the squadron converted to Shackleton MR.2s, followed by MR.3s before returning to Kinloss on 1 April 1959. No. 217 Squadron was a resident at Kinloss from April 1952 with

Left: No 201 Squadron moved from St Mawgan to Kinloss with Shackleton MR.3s on 1 July 1965 and began to replace them with Nimrod MR.1s on 25 June 1970. Shackleton MR.3 XF707 P:201 Squadron on patrol over the sea. *Avro*

Neptunes, but disbanded on 31 March 1957. Other Shackleton units were 201 Squadron which formed at St Mawgan in October 1958 with Shackleton MR.3s and moved to Kinloss on 1 July 1965. It was followed by 206 Squadron which arrived from St Mawgan on 1 July 1965, also flying Shackleton MR.3s. On 8 January 1972 8 Squadron re-formed at Kinloss with Shackleton AEW.2s, but moved to Lossiemouth on 17 August 1973, when improvements had been made to the station. Meanwhile for the operation of more modern aircraft, improvements were made at Kinloss, including the extension of the main runway, the closure of the two other runways, and the construction of large new hardstandings to accommodate the Nimrod Wing consisting of three squadrons.

Kinloss was the home of the Nimrod for some 40 years until its premature withdrawal in March 2010. With the World War 2-vintage Shackleton needing replacement by the late 1960s, Air Staff Target 357 was issued in July 1963 for a sophisticated anti-submarine and coastal patrol jet-powered aircraft. The transit speed was to be at least 400 knots with a patrol time of about eight hours. Various designs were considered, based on a number of current airliners and military transports, but the costs of developing this type of aircraft were estimated to be too high. On 19 June 1964, AST.381 was issued, specifying a less capable aircraft, but closely matching the capabilities of the Breguet Atlantic. As a result the Avro design team proposed their HS.801 based on the well-proven Comet 4 jet airliner, powered by four Rolls-Royce RB.168 Spey turbofans. The major external change was the fitting of a low-drag under-fuselage pannier which housed the surveillance radar, large weapons bay and additional equipment. The initial radar fitted was the nose-mounted air-to-surface vessel ASV-21 system, with provision to replace it by the Searchwater radar which was in development. The under-fuselage pannier was unpressurised and extended to just aft of the wing trailing edge, giving the fuselage a double-bubble section. This project was offered to the RAF in July 1964, and in February 1965 an announcement was made in Parliament selecting the HS.801 to replace the Shackletons. A fixed-price contract for 38 Nimrods, as the type had been named, was agreed in January 1966, and two unsold Comet 4Cs on the production line at Hawarden near Chester were converted to development aircraft, one retaining its Avon jet engines and the other powered by Speys.

Because of the additional side area of the fuselage created by the pannier, the Nimrod fin area was increased by fitting a dorsal fairing and an Electronic Surveillance Measures (ESM)

Above: Kinloss Wing Nimrod MR.2P XV241 patrolling Britain's coastline and protecting our oil fields. May Nimrod RIP. *BAE Systems*

aerial fairing was added to the top of the fin. A 6ft section was taken out of the fuselage forward of the wing to further improve directional stability and a long tail-boom was added to house the Magnetic Anomaly Detector (MAD) to detect the metal masses of submerged submarines. This was located as far as possible from the metallic mass of the aircraft, and was ineffective on the Shackletons, so was never fitted. Visibility for the flight deck crew was improved by deepening the windscreens and 'eyebrow' windows to assist when flying in tight turns over maritime targets. Two wing ribs were strengthened for the carriage of weapons pylons and a 70 million candle power searchlight was installed in the front of the starboard wing pinion fuel tank to assist in search and rescue duties. Known as the Nimrod MR.1, it had 14 crew members, consisting of two pilots, a flight engineer, navigator, tactical navigator, radio operator, radar operator, two sonar operators, an ECM operator, a MAD operator, two surveillance members and a load master. For anti-submarine tasks up to nine torpedoes could be carried, or for use against surface targets free-fall iron bombs could be released. The ASV.21 radar could detect large ships out to 150 nautical miles, surfaced submarines at 75 nautical miles and 'snorting' submarines at 20 nautical miles.

The first production Nimrod MR.1 made its maiden flight from Woodford on 28 June 1968 with five operational squadrons and an OCU to be equipped and based at Kinloss, St Mawgan and Malta. The first MR.1 to enter service was XV230 with MOTU at St Mawgan on 2 October 1969, and a

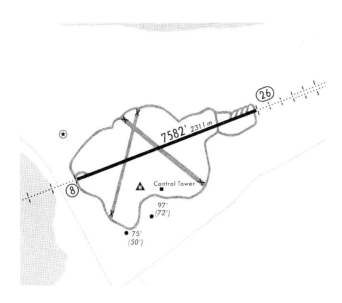

Left: Kinloss was built in 1938/39 to a reduced-specification RAF Expansion standard with three 'C' Type hangars within a technical site located on the southern side of the landing ground. In 1940 45 MU was formed and had a total of 12 "L" Type hangars dispersed around the airfield with the main base on the northern boundary. During World War 2 Kinloss was used for Bomber crew training and three concrete runways were constructed in 1942. Following the end of World War 2, the station took over the maritime reconnaissance role from December 1950, and the first Shackletons arrived in March 1951. The main runway was extended in 1973 for the operation of Nimrods and the other two runways disused. New hardstandings were constructed for the use of three squadrons, the first of which arrived in June 1970. After nearly 40 years of Nimrod operations, the MR.2s were prematurely withdrawn in March 2010 as an economy measure, and the Nimrod MR.4 was cancelled in October 2010 with the closure of RAF Kinloss by March 2013. This removed Britain's maritime reconnaissance capability. The attached approach chart shows the runway layout of Kinloss in September 1967 before the main runway was extended. *Jeppesen*

further eight Nimrods were ordered in January 1972, although only five entered service as MR.1s, one being completed as a MR.2 and the other two retained for conversion to the AEW.3. In addition three special Nimrod R.1s were built for classified electronic-surveillance duties with 51 Squadron.

On 25 June 1970 201 Squadron began to convert to Nimrod MR.1s as the first unit at Kinloss, with 120 Squadron converting to the new jet maritime patrol aircraft in October 1970, and as the aircraft became available 206 Squadron converted in July 1972 on to what was to become known as the 'Mighty Hunter'. The Nimrod MR.1 had always been considered an interim solution pending the planned update of 35 Nimrod MR.1s to the MR.2 standard with Searchwater radar. The conversions started in 1975, 35 aircraft being scheduled for the upgrade with the MR.1s returning to Woodford for the 11-month programme which consisted of seven aircraft out of service at any one time. In addition to the new more capable radar, many of the systems were updated with the first conversion delivered to 201 Squadron on 23 August 1979, the final MR.1 leaving Kinloss in October 1982, and working up on the MR.2 commenced in June 1983. In Britain the Nimrod MR.2 was responsible for anti-submarine warfare, anti-surface unit warfare and search and rescue. The MR.2 could carry up to 25 people, but the normal operating crew of 13 was slightly different from the MR.1. On the flight deck were two pilots and a flight engineer. In the cabin there were two navigators who alternated between routine and tactical roles every other sortie, an AEO who was both sensor and communications co-ordinator. The sensor team consisted of three AEOs ('wet men') who were responsible for monitoring both active and passive sonobuoys and four AEOs ('dry men') who managed the avionics and weapon systems to make the Nimrod most effective.

The Nimrods became part of a Central Servicing system, so that aircraft were made available to the individual squadrons as required from a pool of aircraft, the Nimrods no longer carrying the squadron identities. With the defence cuts of 1992 all Nimrod operations were centred at Kinloss with four Nimrods withdrawn on 1 October and put in storage, reducing the total force to 26 aircraft. In further defence cuts in July 2004 the Nimrod fleet was reduced to 16 aircraft, with some of the surplus allocated to conversion to MR.4s. No 206 Squadron disbanded on 1 April 2005 and 236 OCU, which had been MOTU (Maritime Operational Training Unit) between 1956 and 1965, became 42 (R) Squadron to continue the training role, but without its own dedicated aircraft. With the withdrawal of the RAF Nimrods in March 2010, and the cancellation of the MR.4 in October 2010, RAF Kinloss is due to close to flying on 31 July 2011, with closure as an RAF station on 31 March 2013 and a return to Defence Estates by 31 March 2014. Kinloss is likely to be allocated for Army units returning from Germany around 2015.

St Mawgan, Cornwall;
SW870646, 4 miles E of Newquay.

Aviation first came to the St Mawgan area in August 1933 when Sir Alan Cobham's 'Flying Circus' used a large field at Trebelzue on the cliff tops above Watergate Bay, and it was established for regular commercial flying by Western Airways in May 1939. With the start of World War 2, the site was requisitioned by the Air Ministry to be developed as a satellite to nearby St Eval and opened as RAF Trebelzue in September 1941 with two tarmac runways. The airfield was taken over by Ferry Command in December, but the runways were found to be too short for large aircraft, and work started in August 1942 on the construction of a new three-runway airfield to the east, but adjacent to the original site, which remained operational.

Left: The Shackleton MR.1 was Britain's first dedicated long-range maritime reconnaissance landplane, derived by Avro from the Lancaster/Lincoln designs, but with a wider fuselage. *Author*

The technical site was located on the south side of the new airfield and, being constructed during wartime, was fairly basic with two B.1 hangars and temporary accommodation.

The first operational aircraft were Mustangs of 400 (RCAF) Squadron which arrived in December 1942 with the task of escorting aircraft flying over the Bay of Biscay to Gibraltar, before leaving a month later for Middle Wallop. On 24 February 1943, the station was officially renamed St Mawgan. In June 1943 the station was taken over by the Americans and improvements were made to the main runway ready for the first arrivals on 1 July of USAAF B-17s and B-24s destined for operations in North Africa. There were so many aircraft deployed at St Mawgan that the runways of Trebelzue were used for dispersals with new hardstandings in addition to concrete dispersals located around the perimeter track. During August, September and October large numbers of USAAF aircraft passed through the base, making it one of the busiest airfields in Britain, in addition to commercial use of the airfield by BOAC and KLM. During the spring of 1944 arrivals again peaked in preparation for the invasion of Europe in June, with 169 aircraft arriving from the USA during one day in February. Even with all these movements, construction work continued at St Mawgan with the main runway extended, more hardstandings provided, two T.2 hangars added plus a new control tower and operations room. The extensive airfield facilities were used for weather diversions by a number of RAF and USAAF bombers during early 1945 and with the defeat of Germany, St Mawgan became a key base, initially for the departure of aircraft reinforcements to Asia, and also the return of USAAF B-17s and B-24s to the USA. This included the mass return of 165 Lancasters of 6 Group RCAF to Canada in June 1945. Traffic reduced considerably in 1946, with the station closing on 1 July 1947, and was put under C&M, until it was reopened in January 1951 under the control of Coastal Command.

The Avro Shackleton was Britain's first long-range land-based dedicated maritime reconnaissance aircraft, previous types being the Sunderland flying boats or adaptations of bombers such as Lancasters, Liberators, or the American Hudsons and Neptunes. Following the end of World War 2, the RAF was considering initially the use of Lancasters adapted for the maritime reconnaissance role, to be followed by an adapted Lincoln. However, neither of these types were suitable and on 21 March 1946, Avro received an initial order to Specification 42/46 for 29 airframes for OR.320 for a long-range maritime reconnaissance aircraft. The new aircraft used the basic Lincoln wing married to an entirely new, wider fuselage with a massive weapons bay,

Left: No 42 Squadron operated Shackleton MR.2s at St Eval from January 1953, moving to St Mawgan on 8 October 1958, and began the conversion to Shackleton MR.3s in November 1965. Shackleton MR.2 WL800 A:42 Squadron away from its home base which certainly did not feature 'C' Type hangars. *RAF Museum* P021286

Right: This aerial view of St Mawgan c1960 features the main runway with the World War 2 dispersals around the perimeter track. Shackletons are parked on the apron at the end of the short disused runway with the very basic RAF technical site and accommodation beyond. The original airfield of Trebelzue is out of the picture at the end of the diagonal cross runway.
RAF Museum P017587

accommodation for ten crew and all the necessary weapons and sensors. Power came from four Rolls-Royce Griffon engines driving six-blade contra-rotating propellers. When the aircraft, known as the Type 696, was adopted for service, it became the Shackleton GR.1, named after the famed Antarctic explorer. Planned performance for the new type was a top speed of 300 mph, a service ceiling of 25,700 feet and a maximum range of 3,800 miles.

Avro received the first contract on 28 May 1947 for three prototypes with an ASV Mk.13 radar mounted under the nose giving a range in good conditions of up to 40 miles. The location was not ideal as it did not provide a full 360 degrees coverage. The ten crew members consisted of two pilots, two navigators, a flight engineer and five other members who covered the roles of gunners, signallers, bomb aiming and observation. The bomb bay could carry up to 20,000lb of equipment and ordnance including anti-submarine bombs, depth charges, mines, sonobuoys and markers. Because of the very long endurance required, the plan was to provide reasonable comfort for the crew, including galleys and rest areas. The first prototype Shackleton, VW126, made its maiden flight from Woodford on 9 March 1949. Avro received a contract for the initial 30 Shackleton MR.1s in March 1946 and the first production aircraft made its maiden flight from Woodford on 28 March 1950.

The plan was to base the Shackletons along the western seaboard from St Eval in Cornwall to Aldergrove and Ballykelly, both in Northern Ireland. The first aircraft were delivered to Kinloss, the initial units being 236 OCU and 120 Squadron, its first Shackleton arriving on 3 April 1951 before moving to Aldergrove in April 1952. No. 220 Squadron re-formed at Kinloss in September 1951, moving to St Eval in November, where 269 Squadron re-formed on the type in January 1952, moving to Ballykelly in March. No 240 Squadron re-formed at Aldergrove on 1 May 1951, moving to Ballykelly in June, and 42 Squadron re-formed at St Eval on 28 June 1952. Finally 224 Squadron was re-formed at St Eval on 27 September 1952. Although the intention was to provide soundproofed interiors, this was not done on the earlier marks, the interior of which were painted matt black, resulting in a very noisy and depressing environment. When the early Shackletons were in development, it was understood that with the advances in technology they would soon be out of date, so the Mk.1s were finally replaced by the Mk.2s in February 1962, while 17 of the Mk.1s were converted to the training role as T.Mk.4s for use by MOTU at Kinloss. The T.4s were later replaced by ten conversions of the MR.2s, designated T.Mk.2s.

The development of the Shackleton MR.2 was given the go-ahead on 3 July 1950 and featured a longer nose and rear fuselage with the radar relocated ventrally, aft of the wing, where when lowered to the operating position, provided an all-round coverage. The first aerodynamic MR.2 made its maiden flight on 19 July 1951, followed by the first fully-equipped example on 17 June 1952, both being conversions from the Mk.1s. A total of 70 MR.2s were built and the first to be issued to a squadron was allocated to 42 Squadron in January 1953, followed by 206 Squadron, both at St Eval, in the following month. No 228 Squadron at St Eval re-equipped with MR.2s in July 1954, moving to St Mawgan on 19 November 1956. Despite systems improvements in the Mk.2s, the working environment was no better, with high levels of noise and vibration.

The ultimate development was the Shackleton MR.3 to Specification R.5/46 issued on 18 November 1953, the major external change being the installation of a tricycle undercarriage. In addition to aerodynamic changes to the wing, permanent fuel tanks were fitted to the wing tips. Crew

Left: No 206 Squadron moved from St Eval to St Mawgan with Shackleton MR.2s on 14 January 1958 and re-equipped with MR.3s from January 1958, which the squadron continued to operate until October 1970. The squadron moved to Kinloss in August 1970 to begin the conversion to Nimrods. Shackleton MR.3 XF707 C:206 Squadron with the squadron badge on the fin was patrolling off the Cornish coast, still armed with a pair of guns in the nose turret. *IWM RAF T.4755*

comforts were improved with a full galley and pilot vision was improved with a revised flight deck canopy. The entire interior was lined with soundproofing with a brown and cream finish. The crew members were provided with large comfortable seating and the engine exhausts were in sealed units which effectively and quietly deflected the exhaust outlet below the wings. The first MR.3, WR970, made its maiden flight on 2 September 1955, but was lost fatally on 7 December 1956 during stalling trials from Boscombe Down. It was replaced in the flight development programme by the second production example on 28 May 1956. With development completed, 220 Squadron at St Mawgan was the first squadron to take delivery of an MR.3 on 30 August 1957, and was renumbered 201 Squadron in December. They were joined by 206 Squadron on 14 January 1958 while later production versions were issued to 120 Squadron at Aldergrove and 203 Squadron at Ballykelly. The intention had been to build a total of 52 MR.3s, but a defence review resulted in a cutback to 33 aircraft. With improvements in equipment the MR.3s'

weight increased, resulting in the four Griffon 58 engines being incapable of getting the aircraft off the ground in hot and high conditions. As a result, a pair of Viper jet engines were installed in the rear of the outer engine nacelles with an intake scoop door located below the nacelle, and exhausted out of the rear of the nacelle. The Viper 203 provided 2,700lb of thrust for a weight of 549lb, but because of the additional stresses, the wing structure required strengthening. The Phase 3 conversions, as these updates of the MR.3 were known, were capable of dropping nuclear depth charges.

The first regular flying unit after World War 2 at St Mawgan was the Air-Sea Warfare Development Unit (ASWDU) which arrived from Ballykelly on 4 May 1951, flying a mixture of Lancasters, Shackletons, Ansons and Sycamore helicopters engaged in anti-submarine warfare developments. On 23 October 744 NAS arrived from Culdrose as the Navy's ASWDU to work with the RAF until it was disbanded on 31 October 1956. In December 1956 both 220 and 228 Squadrons moved from nearby St Eval with Shackleton MR.2s in preparation for conversion to the MR.3, for which St Eval was inadequate. No 206 Squadron moved into St Mawgan from St Eval on 14 January 1958, exchanging places with 228 Squadron, to become the second MR.3 unit. No 220 Squadron was renumbered 201 Squadron in October 1958 and 42 Squadron arrived from St Eval with MR.2s prior to its closure, 42 Squadron not receiving MR.3s until November 1965. St Mawgan was now the premier 19 Group Coastal Command base with the task of protecting the southwest approaches to Britain and the eastern Atlantic. It was a St Mawgan-based

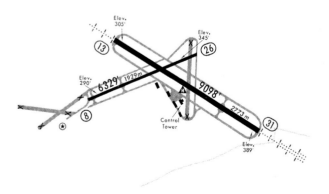

Left: The new airfield at St Mawgan was constructed from August 1942 with three runways to the west of RAF Trebelzue, the disused runways of which are shown beyond the threshold of runway 08. The cross runway was later disused while the main runway was extended to 9,098 feet for Nimrod operations from August 1970. The RAF airfield closed in November 2008, the Nimrods having departed to Kinloss in September 1992, and the site is now operated as an airport by Cornwall County Council. *Jeppesen*

158   UK Airfields of the Cold War

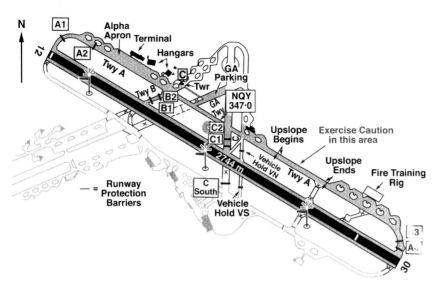

Above: RAF Trebelzue opened for military flying in September 1941 with a very basic technical site on the east side of the flying field, and two runways were constructed. However, these were unsuitable for operations by large aircraft, and a new airfield with three runways was constructed to the north of the original site between August and December 1942. In February 1943 the more appropriate name of St Mawgan was adopted and in June it became a major USAAF transit base. The station was under C&M from July 1947 until January 1951 when Coastal Command took over for the operation of Shackletons, and the main runway was extended from November 1956. The Shackletons were replaced by Nimrods in December 1969, which moved to Kinloss in September 1992. RAF flying ceased in May 2008, but commercial flying which had been going since 1959 continued, the site becoming Newquay Airport. The civil operations are located to the north of the main runway. *Pooley's Flight Guides*

arriving in October to equip the newly formed 236 OCU which began to convert Shackleton crews to the new all jet maritime aircraft, starting with the crews of 42 Squadron from April 1971. On 7 May 1970 7 Squadron was re-formed as a target facilities unit equipped with Canberra TT.18s until it disbanded in January 1982. The Nimrod's main role was to monitor Soviet and Warsaw Pact naval activities, together with ASW training, SAR, fishery protection and oil rig surveillance. Emergencies where the St Mawgan-based Nimrods assisted included the Fastnet Race tragedy in 1979 helping to save many lives. In May 1982 two 42 Squadron Nimrods were deployed to Ascension Island for the Falklands conflict, flying SAR and maritime patrols in the South Atlantic. St Mawgan was also used in May as the departure point for 16 RAF Harrier GR.3s bound for the Falklands. The Nimrods were maintained by a central servicing organisation and shared between 42 Squadron and 236 OCU.

In 1988 the main 9,000ft runway was resurfaced and a group of ten HAS were erected close to the Trebelzue site to allow the station to be used as a forward deployment base for Tornado F.3s. Shortly after these improvements had been made, it was decided to concentrate all maritime reconnaissance capability in Scotland, and both 42 Squadron and 236 OCU were transferred to Kinloss, the last Nimrod departing St Mawgan on 9 September 1992. The station then became home for SAR helicopters and a number of ground-based units.

As early as 1959, commercial flying had returned to St Mawgan with Starways DC-3s operating summer tourist flights from Liverpool, while other operators making use of the airfield as joint users with the RAF. By 1962, passenger numbers had outgrown the modest RAF facilities and a small passenger terminal was built on the northeast side of the airfield close to the village of Carloggas. British Midland brought regular scheduled flights to the airfield in 1969, followed by Brymon Airways which offered regular London-Newquay services. A new terminal was opened in 1993 and extended in 2006 with a number of airlines serving the area.

When it was decided that St Mawgan would not be the home of JSF F-35s in November 2005, plans were announced for closure of the station, the last flying RAF unit departing on 12 May 2008. It closed as an RAF airfield on 30 November 2008, but the site still remains an RAF station as a base for ground units. Cornwall County Council have taken over as owners of the airport with a new air traffic control, new radar, airfield lighting and civil navaids ensuring a commercial future for this fine airfield. A Shackleton MR.2 is preserved within the RAF camp as a reminder of the coastal operations from St Mawgan.

Nimrod which on 30 August 1970 was the first to detect a submerged Soviet nuclear submarine, when a 'November' class hunter-killer was located. In addition to the anti-submarine surveillance, the squadrons were deployed on a number of overseas detachments for operational, humanitarian and goodwill tours. The closure of St Eval confirmed the long-term future of St Mawgan, and efforts were made to replace many of the dilapidated wartime temporary buildings. Changes in 1965 resulted in 201 Squadron going to Kinloss in March followed by 206 Squadron in July, but MOTU came south to join 42 Squadron.

The final Shackleton course took place with MOTU on 28 July 1970 and on 1 August it was renamed 236 OCU to start the crew training on the Nimrod, the first example having been delivered at the end of 1969 to St Mawgan. The three squadrons at Kinloss, 120, 201 and 206, converted to the Nimrod during 1970 after conversion training at St Mawgan. In preparation for the arrival of the first Nimrods, a large new hangar was erected in 1969, the first aircraft

Bibliography and Further Reading

Action Stations Revisited 1 Michael J. F. Bowyer,
   Crecy Publishing ISBN 9 780859 791458

Action Stations Revisited 2 Michael J. F. Bowyer,
   Crecy Publishing ISBN 0 947554 94 7

Action Stations Revisited 3 David W. Lee,
   Crecy Publishing, ISBN 9 780859 791 106

Action Stations Revisited 4 David Berryman,
   Crecy Publishing, ISBN 9 780859 791212

Aircraft of the Cold War Thomas Newdick,
   Amber, ISBN 978 1 906626 63 1

Armageddon, the Illustrated Guide Bob Clarke,
   Amberley, ISBN 978 1 84868 496 6

Boeing B-29 Superfortress Steve Pace,
   Crowood Press, ISBN 1 86126 581 6

Bomber Squadrons of the RAF Philip Moyes, Macdonald

The Buccaneers Air Commodore Graham Pitchfork,
   PSL, ISBN 1 85260 611 8

Canberra – The Operational Record Robert Jackson,
   Airlife, ISBN 1 85310 049 8

English Electric Canberra Beamont & Reed,
   Ian Allan, ISBN 0 7110 1343 8

English Electric P1 Lightning Roland Beamont,
   Ian Allan, ISBN 0 7110 1471 X

Fighter Squadrons of the RAF John Rawlings, Macdonald

Force for Freedom Michael J. F. Bowyer,
   Patrick Stephens, ISBN 1 85260 415 8

Gloster Meteor Chaz Bowyer,
   Ian Allan, ISBN 0 7110 1477 9

Gloster Javelin Maurice Allward,
   Ian Allan, ISBN 0 7110 1323 3

The Harrier Story Davies & Thornborough,
   Naval Inst. ISBN 1 55750 357 5

Hawker Hunter Robert Jackson,
   Ian Allan, ISBN 0 7110 1216 4

History of the RCAF Christopher Shores,
   Royce, Toronto ISBN 0 86124 1606

A History of the USAFE Cees Steijger,
   Airlife ISBN 1 85310 075 7

Jane's All the World's Aircraft, various editions

The last of the Lightnings Ian Black,
   Sutton Publishing, ISBN 0 7509 3073 X

Lincoln at War 1944-1966 Garbett & Goulding,
   Ian Allan ISBN 0 7110 0847 7

Military Aircraft of the Cold War Jim Winchester,
   Grange, ISBN 1 84013 929 3

Military Airfields of Lincs etc Bruce Barrymore Halpenny,
   PSL, ISBN 0 85059 484 7

Military Airfields of Yorks Bruce Barrymore Halpenny,
   PSL ISBN 0 85059 532 0

Military Airfields of Scotland etc David J. Smith,
   PSL, ISBN 1 85260 309 7

The Nimrod Andy Evans,
   Dalrymple & Verdun Publishing ISBN 1 905414 06 2

Panavia Tornado Andy Evans,
   Crowood ISBN 1 86126 201 9

The last of the Phantoms Ian Black,
   Patrick Stephens, ISBN 1 85260 612 6

RAF Aircraft Today – Phantom Bill Gunston,
   Ian Allan, ISBN 0 7110 1383 7

RAF Squadrons Wing Commander C. G. Jefford,
   Airlife, ISBN 1 85310 053 6

Sabre in RAF Service Duncan Curtis,
   Sutton Publishing, ISBN 0 7509 4236 3

SEPECAT Jaguar Andy Evans,
   Crowood, ISBN 1 86126 144 6

SEPECAT Jaguar Arthur Reed,
   Ian Allan, ISBN 0 7110 1215 6

Shackleton Richard A. Franks,
   Dalrymple & Verdun Publishing ISBN 1 905414 01 3

Strike Force Robert Jackson,
   Robson Books ISBN 0 86051 303 3

V-Bombers Robert Jackson, Ian Allan ISBN 0 7110 1100 1

Vickers Valiant Tony Buttler, Warpaint Books

Vickers Valiant Eric B. Morgan, Aerofax, ISBN 1 85780 134 2

Handley Page Victor Andrew Brookes,
   Ian Allan ISBN 0 7110 1803 0

Handley Page Victor Steve Hazell, Warpaint Books

Avro Vulcan Kev Darling, Hall Park, ISSN 1361 0369

Avro Vulcan Kev Darling,
   Speciality Press, ISBN 1 58007 023 X

Avro Vulcan Philip Birtles,
   Ian Allan, ISBN 1 85780 269 1

World War 2 Airfields Philip Birtles,
   Ian Allan, ISBN 0 7110 2681 5